1953 Working in Cambridge, England, James Watson and Francis Crick develop the double-helix model for DNA, where the two helices are joined together by pairs of nucleotides that look like the rungs of a ladder.

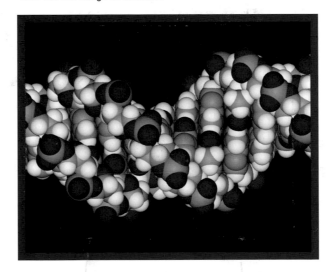

THE ILLUSTRATED TIMELINE OF
Science

THE ILLUSTRATED TIMELINE OF
Science

A CRASH COURSE IN WORDS & PICTURES

Sidney Strickland, PhD, with Eliza Strickland

A JOHN BOSWELL ASSOCIATES BOOK

STERLING PUBLISHING CO., INC.

NEW YORK

Library of Congress Cataloging-in-Publication Data Available

2 4 6 8 10 9 7 5 3 1

Published by Sterling Publishing Co., Inc.
387 Park Avenue South, New York, NY 10016
© 2006 by Sterling Publishing, Co., Inc.
Distributed in Canada by Sterling Publishing
c/o Canadian Manda Group, 165 Dufferin Street
Toronto, Ontario, Canada M6K 3H6
Distributed in the United Kingdom by GMC Distribution Services
Castle Place, 166 High Street, Lewes, East Sussex, England BN7 1XU
Distributed in Australia by Capricorn Link (Australia) Pty. Ltd.
P.O. Box 704, Windsor, NSW 2756, Australia

Sterling ISBN-13: 978-1-4027-3604-9
ISBN-10: 1-4027-3604-5

For information about custom editions, special sales, premium and
corporate purchases, please contact Sterling Special Sales
Department at 800-805-5489 or specialsales@sterlingpub.com.

Book Design by Barbara Aronica-Buck

Contents

Acknowledgments

Thanks are first due to the team at John Boswell Associates—
John Boswell, Christa Bourg, and Lauren Galit—who conceived
the idea for this book, and carefully shepherded it though the long
process to completion. We were also extremely lucky to work with
the designer, Barbara Aronica-Buck, who took a wealth of information
and turned it into an eye-catching, elegant book.

This book is a family effort, and this father-daughter team would
like to thank the other members of the immediate family. Carol devoted
many hours to reading pieces of the manuscript, and never failed to offer
wise and illuminating advice. Alison and Joe have provided unflagging
interest and enthusiasm, and little Lora has been an inspiration—already
her curiosity and keen observational skills show her scientific aptitude.

S.S. thanks his scientific mentors: Priscilla Rushton, Martin Morrison,
Vincent Massey, and Edward Reich. He dedicates this book to Carol,
for thirty-eight years his *sine qua non*.

E.S. thanks Chris Thompson for his dependable delight in discussing such
topics as Tycho Brahe's silver nose and the miasma theory of disease.
Here's to many years of conversation.

Introduction

Why should you care about the history of science? First, there's the easy answer: because it will give you a better understanding of our modern lives. The materials and technologies that we take for granted are the result of a long chain of discoveries made by previous generations of scientists.

But there's also a loftier reason. Learning the story of scientific advancement is a way of honoring some of the nobler impulses in humanity. Scientists are driven by curiosity and a desire to understand the marvelous intricacies of the world. The awe and wonder that we feel when confronted with nature's grand design is one of the most fundamental human emotions. Early cultures across the globe—from Mesopotamia to the Mayans in South America—all gazed up at the heavens and tried to understand why the moon waxed and waned and the stars wheeled above them.

This constant quest for knowledge has brought us far. Having delved into our own genetics and explored the synapses in our brains, we now know more than ever before about our physical makeup. We have explored the depths of the planet we inhabit and have sent robots to skim over the surface of Mars. Yet our curiosity is still a mysterious driver; we can't fathom what makes us try to fathom the world. As the great physicist Richard Feynman wrote in a poem, we are "atoms with consciousness; matter with curiosity."

That consciousness and curiosity has brought our species, one step at a time, from caves in Africa to the surface of the moon. Browsing through the scientific history presented here, you can see that the knowledge gained is cumulative, and can contemplate how each advance built on earlier work. The inventions, theories, and discoveries detailed in these pages are usually attributed to individuals, but they're really the product of our collective, human yearning for understanding.

To those researchers and thinkers who were in the thick of it, however, the march of progress was not clear. When early empires fell, information got lost and techniques were forgotten. Wrong ideas surged forward and dominated for centuries, while many challenging theories were met with outrage and derision. Copernicus first formulated his theory of a heliocentric universe in 1514, refuting the central tenet of scientific understanding that described the sun circling the earth. It took over 100 years for scientists to accept his theory as fact, while the Catholic Church continued to deny it until the mid-1700s. But once enough evidence had accumulated, scientists embraced the discovery and used it as a jumping off point for new work.

Included in this book is a sampling of the great scientific discoveries and advances to date—only a sampling, unfortunately, because the breadth of accomplishment in biology, chemistry, physics, earth science, and astronomy doesn't permit a full accounting. The survey begins over two million years ago, when human ancestors first turned stones into cutting tools. It concludes in these first years of the 21st century, with modern man grappling with an unprecedented power to manipulate the environment, from tiny molecules to sweeping climate systems. The illustrated entries are laid out chronologically, on a timeline, to allow the reader to easily trace the human journey through the centuries and millennia.

How to Read This Book

The 2 million years of history covered in this book are divided into four major sections, each with its own introduction. These introductions provide context for the scientific advancements and discoveries, with information about the socio-political state of the world during each era. They cover the intellectual trends, theological concerns, and political upheavals that influenced scientists and their attempts to understand the world around them.

The individual entries are placed along the central timeline, and the text of each focuses on the work's particular importance in the development of science. The entries vary, from explaining the mechanism of an invention to a succinct statement of a scientist's main theory or contribution to science. Some scientists had an immediate and tangible impact, such as those who invented x-ray machines, new forms of plastic, or new "wonder drugs." Other times, a scientist's main role was to advocate a new way to study the world, like those who called for a scientific method comprised of careful observation and controlled experimentation.

Entry texts are color-coded to tell you, at a glance, where in the world each discovery or advance occurred. Sidebars are scattered throughout the book to more thoroughly explain a topic, or to provide more details about an intellectual trend or an individual scientist. This highly browsable, visual presentation will help you understand and enjoy the long story of scientific development, whether the book serves as an introduction or just a refresher course.

The great French biologist Louis Pasteur called science the torch that illuminates the world. Taking a long view of history, it's easy to see the torch being passed from hand to hand, country to country, as each successive generation built on the advances of the past.

Caption Color Key

▉	= Europe
▉	= North America
▉	= Asia
▉	= Africa
▉	= South America
▉	= Australia
▉	= International

From Stone Tools to Copernicus: 2.4 Million B.C. to 1514 A.D.

The early ancestors of human beings emerged in eastern Africa, where each day was a struggle for survival. What could be called the first scientific experiments were conducted by hairy, heavy-browed hominids who were trying to make their lives a little easier: they chipped away parts of stones to give them sharp edges, and made better cutting tools. Gradually, they learned to manipulate their environment, harnessing natural phenomena like fire and creating the first technologies, such as oil lamps and spear-throwers.

When humans began to cluster together in the first civilizations during the Mesolithic period, around 10,000 years ago, they also began to understand the concept of division of labor—some people hunted, while others cooked the food and tended to the settlement's needs. Some cultures began farming, an efficient means of food production that allowed a new class of scholars time to look up at the sky with wonder. Curiosity about the cosmos was nearly universal among the scattered civilizations of the time. In South America, Mayan astronomers observed the rising and setting of the sun and the moon's phases, and learned to chart these natural cycles. Their studies had practical applications as well; better knowledge of the seasons helped in their early agricultural efforts.

Several thousand years later, the philosophers of ancient Greece began to ponder the next step in their graceful cities of slender columns and silver olive trees. These "natural philosophers" moved beyond practical technologies and rudimentary observations, and searched for the general principles that governed the way the world works. They sought to understand how the earth fits into the solar system, what matter is composed of, and what laws govern bodies in motion.

In China, halfway around the world, educated scholars made great advances, and produced many technologies that weren't seen in Europe until the Middle Ages. The Han Dynasty ran an enormous, rigidly centralized empire for over 400 years: an imperial census in the first century A.D. counted a population of 57 million. Technological innovations helped the empire function smoothly, with canals to improve transport and a nationalized system of iron production. However, their many inventions often didn't spread westward because stable trade routes had not yet been established. It's difficult to know in which cases western thinkers were influenced by accounts they'd heard of the marvels of east, and when they discovered something independently, albeit many centuries later. When the Han empire fell in 220 A.D., progress slowed.

Meanwhile, the Roman empire carried on Greek traditions in many ways, although it emphasized technology over philosophy and abstract thought. Roman engineers built roads, aqueducts, and enormous coliseums that still amaze visitors today. But the light of learning was soon to fade from Western Europe. In 330 A.D., the Roman emperor Constantine dedicated the city of Constantinople as the new imperial capital. Rome and its surrounding territories had been long under attack by Germanic tribes and the Huns, and when the invaders sacked Rome in the 5th century, learning shifted to more stable cities in the east.

Some documents and treatises were brought to Constantinople for safe keeping, while many others landed in thriving Arab cities across the Middle East. In 832, the Arab caliph al-Ma'mum of Baghdad founded the scholarly "House of Wisdom" where the great works of the ancient Greeks were preserved and translated into Arabic. Arab scholars also made independent advances in fields such as optics, mathematics, and chemistry. The Islamic state began to break apart in the 900s, as rival dynasties struggled for power, but the intellectual centers continued to produce important work for several centuries.

Europe's medieval period (c. 475 to c. 1450) was characterized by the Church's dominance over and resistance to science. Christianity emphasized the glories of the afterlife, and Church leaders viewed the scientific drive to understand the physical world as a dangerous and impious impulse. When the works of Aristotle and many other important Greek writers began to be translated from Arabic to Latin in the 1100s, Europe's scholars embraced the forgotten works, but also sought to reconcile them with Christian doctrine. This philosophy of scholasticism dominated European thought, wherein philosophers laid down previously accepted tenets and then attempted to bring new observations into accord with those old ideas.

A few European thinkers bucked the trend, though often at considerable expense. The Franciscan friar Roger Bacon (c. 1214–1294) grew frustrated with the limits of scholasticism, and emphasized the importance of experiments instead of the blind acceptance of wisdom from the past. His ideas met with plenty of resistance: some claim he was imprisoned by the Franciscan order for his experiments in alchemy and his statements about the ignorance of the clergy.

Three centuries later, the Polish astronomer Nicolaus Copernicus followed in Bacon's footsteps, taking a fresh look at conventional wisdom and cherished beliefs. By 1514 he had already sketched out his idea of a heliocentric solar system, in which the earth and other planets revolve around the sun. This idea rebutted the accepted Church doctrine of an earth-centered solar system, with God's hand guiding the sun, moon, and planets in perfect circles around the earth. Copernicus completed his masterwork, *On the Revolution of Celestial Bodies*, in the 1530s but didn't publish it until just before his death, fearing reprisals from the Church.

Indeed, the Church soon placed Copernicus's book on the list of prohibited texts, but his ideas continue to spread among astronomers. The Italian Galileo Galilei was the greatest champion of Copernicus's ideas, although the Catholic Inquisition forced him to publicly recant those beliefs. The work of these bold men set the stage for the coming Scientific Revolution, wherein scholars learned to value empiricism rather than doctrine. Galileo gave eloquent voice to this new approach: "In questions of science the authority of a thousand is not worth the humble reasoning of a single individual."

2.4 Million B.C. Early hominids in east Africa create first stone tools. Homo habilis—the "handy man"—chips pieces of stone into the shapes of hand axes, used for cutting meat and scraping hides.

79,000 B.C. Simple forms of stone lamps are used in southern Africa. *Homo sapiens* hollows out pieces of limestone or sandstone, use animal fat for fuel and grass or moss for wicks. Provides light for cave painting and craftwork.

30,000 B.C. Cro-Magnons in North Africa and Europe invent the bow and arrow. Use chipped and flaked pieces of flint for arrowheads. Increases efficiency in hunting and life expectancy of hunters.

1,000,000 B.C.

1 Million B.C. *Homo erectus* in southern Africa learns to make fire. Previously, the nomadic people looked for fires kindled by lightning, and tried to keep them burning. Benefits of their new ability: consistently cooked food, better health, more mobility, and more social interactions as they gather around fires at night.

Early Humanity

The earliest hominids distinguish themselves from their primate ancestors by standing upright on two feet. When they learn to stride over the savannahs of eastern Africa, their hands are free for other uses, like carrying supplies or hunting equipment. Eventually they put their hands to use making tools, marking the first step in human cultural development.

9000–8000 B.C.
Wheat and barley cultivated in Canaan (Israel), the beginning of agriculture. Enables the switch from a migratory, hunter-and-gatherer lifestyle to settled civilization.

8000–7000 B.C.
Mesopotamians use clay tokens (much like those pictured from the 4th century B.C.) to represent a certain number of animals or quantity of grain in trading transactions. A first step towards the use of numbers.

20,000 B.C. People in the south of France and Russia make sewing needles out of bone and antler. They make string out of animal gut, and sew together pieces of softened leather.

9000–8000 B.C.
The Maya make astronomical inscriptions and build observatories in Central America to track and record the movements of the moon, Venus, and other celestial bodies. Some ceremonial buildings are precisely aligned with the sun's path on the fall and spring equinoxes.

6000 B.C. Sumerians begin to produce beer, which they drink together from a communal bowl. This controlled fermentation of barley is the first example of biological engineering. A hymn to Ninkasi, the Sumerian goddess of brewing, explains the procedure: loaves of bread are mashed with malted barley and honey, and fermented in jars.

Sowing the Seeds

The switch from hunting and gathering to agriculture allows humans to establish the first permanent settlements. Around 5500 B.C. the Sumerians begin large-scale agricultural production; they grow wheat and barley, and use an irrigation system and a specialized labor force. Increased production allows farmers to feed large numbers of city-dwellers who then don't need to spend their time in the fields, a crucial factor in the rise of standing armies. Agriculturalism of the Sumerians allows them to embark on an unprecedented territorial expansion, making them the first empire builders.

6000–5000 B.C. Chinchorro Indians in coastal Chile and Peru mummify their dead, more than two thousand years before the Egyptians (pictured: mummy of Ankhef, 11th Dynasty). Show knowledge of anatomy by disassembling bodies, removing soft tissues and organs before stuffing skin with vegetable matter to fill out the human shape.

5000–4000 B.C. Mesopotamians build sailing ships with a single square sail (similar to this model from 12th Dynasty, Egypt), using tar to waterproof the crafts. Ships allow them to venture beyond the rivers to open sea to become the first traders in the Persian Gulf.

4000–3500 B.C. Egyptians create the sundial to tell time, with markings for the hours. Eventually adjusted to give accurate readings throughout the year, despite seasonal changes. A refinement of the previous technique for telling time, where a stick in the ground cast a shadow.

3500–3000 B.C. Sumerians invent wheeled vehicles with wheels made of wooden disks, and begin using 4-wheeled chariots in battle. Horse-like creatures called onagers pull woven carts where two men stand upright. Vehicles are heavy and hard to maneuver, but convey status and strike fear in the enemy ranks.

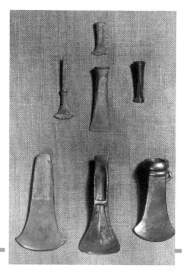

3500–3000 B.C. Egyptians and Babylonians use bronze, a mixture of copper and tin. It's stronger and harder than copper, suitable for weapons, armor, jewelry and tools. Dominant metal for about two thousand years.

5000–4000 B.C. Egyptians mine copper ores from the Sinai peninsula and refine copper by heating the minerals in a charcoal fire. The most advanced metallurgy of the time. Use copper for plates, tools, jewelry, and eventually for drainage pipes in their early plumbing system.

The Sumerians

Along the banks of the fertile Tigris and Euphrates rivers in Mesopotamia, the Sumerian people build Uruk, which is thought to be the first city-state. They expand outward from there, and by about 2500 B.C., their empire stretches across the top of the Arabian peninsula, from the Mediterranean to the Persian Gulf. The densely populated cities—like the city of Ur, with its population of 24,000—give rise to commerce, trade, and encourage the development of practical technologies.

c. 3000 B.C. King Menes of Egypt dams the Nile to collect water from torrential rains and convert marshes to farmland. The earthen dam creates a reservoir, from which farmers irrigate their fields with short canals.

2635–2595 B.C. Imhotep is the founder of Egyptian medicine and the first physician known to written history. Describes human anatomy and cures for over 200 ailments, treating diseases like arthritis, gallstones, appendicitis, and tuberculosis with medicinal herbs and surgery. Later worshiped as the god of medicine by the Egyptians.

The Metal Ages

Early advances in civilization bring about increasingly sophisticated metallurgy, and so we often refer to epochs by the metal in use at the time. As the Copper Age (c. 7000 B.C. to 3500 B.C.) gives way to the Bronze Age (c. 3500 B.C. to 1600 B.C.), better tools create more efficient societies, and better weapons allow those societies to dominate their neighbors. The Iron Age is delayed because processing iron requires high furnace temperatures only achievable with bellows, which won't be invented until around 1600 B.C.

3000 B.C.

365 Days

2773 B.C. Egyptian calendar of 365 days created, an improvement on the previous calendar of 360 days that routinely fell out of sync with natural phenomena. New calendar has 12 months of 30 days, and 5 days of festival when the Nile floods.

c. 2580 B.C. The Great Pyramid of Giza is completed as a burial vault for the pharaoh Khufu, also known as Cheops. An engineering marvel, requiring over 2 million blocks of stone to be cut to precise measurements, some of which weigh 15 tons. Demonstrates geographic knowledge; the pyramid is oriented with its sides along exact north-south and east-west axes.

2500–2000 B.C. Egyptians begin to use papyrus as a material to write on. Criss-crossed layers of the reeds are soaked, beaten until they stick together, then dried in the sun. Use of papyrus improves record keeping, and allows ideas to circulate more easily.

2200–2100 B.C. Stonehenge finished, probably used as an observatory and for sun-worshiping rites. Stone entryways frame the rising sun on the day of the summer solstice and the setting sun of the winter solstice. Stones are transported 240 miles from the southern tip of Wales, probably using sledges and rafts.

2296 B.C. Chinese observers record the passing of a comet, the earliest known record of a comet sighting.

Mesopotamians and Math

Humans had used tally marks for some time—simple slashes to indicate numbers—but Sumerians invent the first number system around 3000 B.C. This first place-value system—based on 60 instead of on 10—survives today in our measurements of time and angles. Mesopotamians also develop the abacus, create a multiplication table, solve equations, and discover what is now known as the Pythagorean theorem, which describes the relationships among the lengths of the sides of a right triangle ($a^2 + b^2 = c^2$).

2000–1950 B.C. First rough plows in use in Uruk (Iraq) and Canaan (Israel)—a frame holds a vertical wooden wedge that is dragged through the topsoil. The first true plows made of bronze are used in Vietnam a few centuries later.

1400 B.C. The Hittites begin the large-scale production of iron, using a blast furnace to facilitate melting. Iron weapons give them an advantage in warfare, leading to the rise of the Hittite empire, which stretches across Mesopotamia and Anatolia (Turkey).

1450–1400 B.C. Water clocks, used in Egypt, measure time by the amount of water flowing from one container to another through a tiny aperture. Greeks use them a millennium later, naming them *clepsydras*, which means "water thieves."

Ancient Egypt

Around 3000 B.C., the Egyptians begin building their cities along the Nile River. As their empire expands over the next millennium, traders bring back raw materials like copper, gold, and cedar, which leads to the flourishing of crafts and technology. The fame of Egyptian healers arises from the religious practice of mummification, which gives the physicians an intimate knowledge of anatomy.

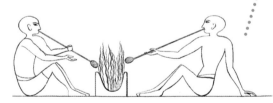

1400–1350 B.C. The Egyptians and Mesopotamians produce glass containers, melting sand and shaping it into containers around a clay core. Previously, glass was used in Egypt to glaze surfaces and for beads and mosaics.

c. 1200 B.C. Ducts beneath the floors of the King of Arzawa's palace in Anatolia (Turkey) suggest that it had central heating. No other evidence of the practice for the next thousand years. (Pictured is a later example of floor heating found at ruins in Epheus, Turkey.)

c. 800 B.C. Liquor is distilled in parts of Asia, showing the ability to manipulate liquids and gases. Fermented beverages are heated until the alcohol begins to evaporate, then vapors are collected and condensed. Arab alchemist Geber discovers distillation anew in 8th century, but considers alcohol an interesting chemical product rather than a beverage.

1200 B.C.

c. 1000 B.C. Chinese develop several food preservation techniques, including salting, drying and smoking, and fermentation in wine (vinegar).

Greeks Search for General Principles

The Greeks establish institutions of learning and research such as the Academy, the Lyceum and the Museum. They look for general theories to explain earthly and celestial phenomena, rather than relying on supernatural causes. Greek philosophers question the shape of the earth and the structure of the solar system, and wonder about the fundamental nature of matter: some believe all matter is a form of water or air. While most of their ideas are wrong, their embrace of inquiry and scientific debate allows many theories to be proposed.

c. 700 B.C. Engineers construct a limestone aqueduct to bring water to the Assyrian capital city of Nineveh (pictured in this carving from c. 645 B.C.). Where it crosses a valley near the city, it's 30 feet high and 900 feet long. Romans improve on the design several centuries later.

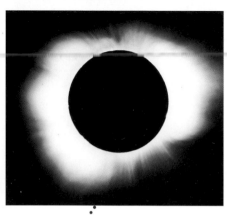

585 B.C. Thales of Miletus (Turkey) reputedly predicts a solar eclipse that occurs on May 28. Two warring kingdoms, the Medes and Lydians, take the eclipse as a bad omen and call off a battle. Thales also studies astrology and looks for natural explanations for phenomena like earthquakes and lightning, strongly influencing Aristotle.

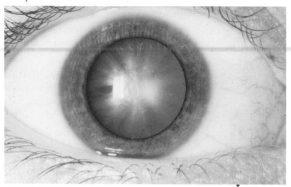

500–450 B.C. Indian surgeon Susrata performs the earliest documented cataract operations. Cuts through the eye's protective layer and pushes back the part of the lens clouded by the cataract.

c. 500 B.C. Chinese discover natural gas wells, use the flammable gas to evaporate brine and make salt. Later make methane lamps by filling a leather bladder with gas from a well, making a small slit and lighting the escaping gas.

PYTHAGORAS PHILOSOPHE
Grec. Chap. 25.

500–490 B.C. A group of Greek scholars called the Pythagoreans teach that the earth is a sphere, not a flat disk. Led by Pythagoras, the scholars also propose that the earth spins on its axis, a simpler explanation than stars, sun, moon, and planets all revolving around the earth.

c. 486 B.C. Chinese engineers begin constructing a canal linking the Yangtze (pictured) and Huai rivers. Later, in 605 A.D., this canal and others are incorporated into the Grand Canal, the largest artificial river in the world.

400–390 B.C. Hippocrates founds the profession of physicians in Greece, encourages the separation of medicine from religion. Says disease results from natural rather than supernatural causes. History attributes the medical code of ethics known as the Hippocratic Oath to him, as well as the aphorism, "First, do no harm."

Aristotle

Of all the ancient Greek scholars, Aristotle's ideas have the most influence on science; many of his methods and theories have been accepted as doctrine for over a millennium. He believes that humans can elucidate all the laws of the universe using reason and the study of previously accepted tenets, without open-minded observation and experimentation. While the reliance on this method slows scientific progress, Aristotle is also remembered for making great advances in the life sciences.

500 B.C. The Greek physician Alcmaeon of Croton, a disciple of Pythagoras, is the first known person to dissect cadavers for science. Recognizes the brain as the seat of intellect. Says that sense organs are connected to the brain, and notes the optic nerve.

350–340 B.C. Aristotle classifies over 500 species of animals thanks in part to his pupil, Alexander the Great, who sends samples of plants and animals from all corners of the empire. Also studies embryology and shows that women play a role in creating the embryo, rather than merely housing it in the womb.

320–300 B.C. A pupil of Aristotle's named Theophrastus publishes the first two treatises on botany. The works are the authoritative texts on the subject for the next 1,500 years. Explains things like seed germination, grafting, pollination of date palms and the sexuality of plants.

300–290 B.C. First clear reference to a lodestone's alignment with the earth's magnetic field in a Chinese text: called a "south-pointer." Lodestone, also known as magnetite, is a magnetic mineral, a naturally occurring form of iron oxide.

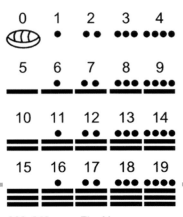

Women in Ancient Greece

The Pythagorean school and Plato's Academy admit women as students and teachers, but otherwise there are not many opportunities for education. In medicine, women are mostly restricted to the role of folk healers who concoct herbal remedies. However, one Greek woman, Agnodike, disguises herself as a man in order to train as a physician, and goes on to treat the women of Athens.

300–290 B.C. Euclid's "Elements" organizes math knowledge developed in Greece over 3 centuries, includes info on plane and solid geometry. The book becomes the basic textbook in math for the next 2000 years. Starts by defining basic concepts—like a point, a line, a surface, a circle—and then sets forth a series of theorems based on these starting points.

260–240 B.C. The Mayans use a place-value number system based on 20, which is soon altered to include a zero. Most sophisticated system used until 600 years later, when Indian mathematicians independently begin to work out a numeral system that uses the zero symbol.

14

300–250 B.C. Steel produced in India by heating high-quality iron with charcoal, a process Europeans would call the "crucible technique" many centuries later. The melted iron absorbs the carbon, and can be forged into exceptionally hard weapons.

c. 200 B.C. Chinese note the six-sided, hexagonal nature of the snowflake. Not understood in Europe until the late 16th century.

260–240 B.C. Archimedes, a Greek mathematician and inventor, works out the principle of the lever. Demonstrates it by pulling a fully-laden ship onto land single-handedly, using a system of compound levers in pulley form.

240 B.C. Eratosthenes, the director of the library in Alexandria, Egypt, estimates the earth's radius by comparing shadows at different locations. His estimate is fairly accurate, within a few percent of the correct value.

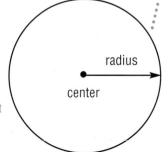

200–190 B.C. Builders in the Roman town of Palestrina (Italy) use concrete—a mixture of sand and small stones bound together by volcanic ash that hardens when mixed with water.

100 B.C. Greeks devise the earliest type of astrolabe, used to find and predict the positions of the sun and stars. Can be set to show the appearance of the sky for any date or time. As developed further by Arab scholars centuries later (pictured), they're used to determine latitude and to study the movements of celestial objects.

132 A.D. First seismograph invented by Zhang Heng in China. Indicates the direction of an

earthquake by dropping a ball from the mouth of a bronze dragon to the mouth of a bronze frog.

100 A.D.

40–90 A.D. Pedanius Dioscorides, a surgeon who travels with the Roman army, compiles a book on medicinal herbs. He describes uses for cinnamon, ginger, wormwood, juniper, and lavender, among many others. Book remains in use until the 1600s.

100 A.D. Chinese use crushed chrysanthemum flowers to kill insects, first insecticide. The active ingredient, pyrethrum, is still used today.

Chinese Technology

Under the stable government of the Han dynasty (206 B.C. to 220 A.D.) the Chinese make great advances in the applied sciences, inventing devices that aren't seen in Europe for a millennium. The gentry's emphasis on education combines with a strong internal trade system to encourage the flow of ideas. In an early agricultural advancement, they invent the collar harness for horses in 120 B.C., which are not used in the West until the Middle Ages. Other early inventions include the crank handle for turning wheels, the ship's rudder, woven silk, cast iron bridges, and, in 105 A.D., the first use of paper as a writing material.

170 A.D.
Roman physician and anatomist Galen of Pergamum (Turkey) uses the pulse as a diagnostic tool. His writings on medicine are

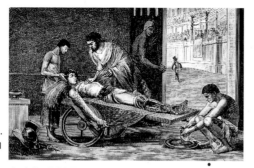

used until the end of the Middle Ages. Theorizes that the body contains 4 vital fluids—blood, phlegm, yellow bile and black bile—and that diseases are caused by an imbalance of the fluids, or "humours."

Alchemy in Alexandria

Through the practice of alchemy, a precursor to modern chemistry, researchers seek the secret of transmutation—transforming common metals into silver and gold. While it is a mystical and misinformed pursuit, alchemists do spur a new interest in experimentation, and discover many basic laboratory techniques that are in use today, including distillation, evaporation, and filtration. Many early Egyptian alchemists are women who have experience in making perfumes and cosmetics.

C. 370 A.D. Hypatia, a teacher at the Museum of Alexandria (Egypt), is the best known female philosopher, mathematician, and astronomer in ancient times. A pagan, she is murdered by fanatical Christian monks.

140 A.D. Ptolemy, the last great astronomer of the Alexandrian school in Egypt, writes that the Earth is the center of the universe, with sun, moon, and stars orbiting in perfect circles. Accepted theory until Copernicus, 1400 years later.

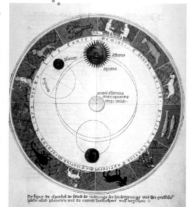

250–300 A.D. In his 8-book masterwork, "Mathematical Collection," Pappus of Alexandria describes five machines in use: cogwheel, lever, pulley (an example of one is pictured), screw, and wedge. Also summarizes the state of mathematical knowledge, and proposes geometric theorems. One of the last important ancient Greek scholars.

c. 400 Zero symbol first recorded in India. Called "sunya," meaning void, or empty. May have been invented centuries earlier, when Buddhists developed the philosophy of "shunyata" or zero-ness. An important advance in the Hindu-Arabic numeral system used today.

0

The Dark Ages

The early medieval years are often referred to as "the dark ages" because the expansive Roman empire disintegrates under a wave of invasions, causing the centers of learning to decline. In addition, an early, fierce version of Christianity puts more emphasis on religious faith than scientific inquiry. The library at Alexandria burns in 269 and 389, and the Academy and Lyceum in Athens are closed by the Byzantine emperor in 529, who views their teachings as pagan. Scientific progress in Europe slows until the mid-15th century, when the fall of Constantinople brings fleeing scholars and their manuscript collections to Europe.

532 The Byzantine emperor Justinian orders the Hagia Sophia cathedral built in Constantinople. The dome is 120 feet wide at its base, and 46 feet high. Weight of the dome is supported by four massive stone arches. Shows advanced engineering knowledge.

c. 750 The Arab alchemist known as Geber advocates experimentation and is later known as the father of chemistry. Invents laboratory equipment, discovers chemical processes like crystallization, and describes how to prepare chemical compounds. Helps change chemistry from a mystic set of rituals to a rigorous science. (Pictured: title page from Geber's *Philosophy of Alchemy*).

600

-4.75

c. 600 Decimal notations used in Indian mathematics. Soon after, the Indian mathematician Brahmagupta writes a book of trigonometry and algebra that states the rules for arithmetic on negative numbers and zero.

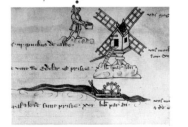

c. 650 First windmills built in Persia (Iran). First used to pump water, soon after adapted to grind grain by turning a grinding stone. Put into use in England and France in 1100s.

c. 740 A Buddhist charm scroll is the earliest known printed text, created using woodblocks. Soon after, the first newspaper is printed in Beijing.

c. 1025
The Arab physician al-Haytham, known to the west as Alhazen, writes the "Treasury of Optics." Describes the structure of the eye, how lenses work, and how to construct parabolic mirrors that are eventually used in telescopes.

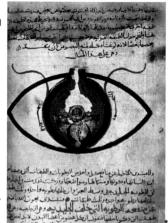

Science in Arab Lands

After the fall of Rome, the wealthy and stable Arab cities become a prime destination for scholars. Around 830, the science library Bayt al-Hikma, or "house of wisdom," is founded in Baghdad. It becomes a center for translation and scholarly research, along with other libraries established in Egypt and Spain. Many ancient texts are preserved through the dark ages by Muslim scholars: for example, the "Astronomical System" of Ptolemy is translated from Greek into Arabic, and centuries later brought to Europe and translated into Latin.

c. 830 Persian mathematician al-Khwarizmi shows a system for solving equations in his book on algebra, a word that comes from part of his book's title: "al-jabr" (a page from which is pictured). Also translates books of Indian mathematics, and encourages the use of the "Indian numerals" that we use today—because Europeans learn of the numbering system from al-Khwarizmi, they mistakenly call them "Arabic numerals."

c. 900 Abu Bakr al-Razi, Persian physician and scholar, writes a multitude of works on medicine. Describes symptoms and treatments of infectious diseases like smallpox and measles, and proposes that fever is the body's way of fighting disease. Becomes chief physician of a Baghdad hospital, where he is the first to use plaster casts and animal gut sutures.

c. 1040 The first recipes for three varieties of gunpowder are published in China by Tseng Kung-Liang. Probably used in bombs, mines, and flame-throwers. Rockets invented soon after, although the "fire arrows" are inaccurate and used primarily to frighten enemies in battle.

19

c. 1060 The renowned female physician Trotula teaches at the first medieval medical establishment not connected with the Church in Salerno, Italy. Writes treatises on conception, childbirth, and infant care, among other topics.

c. 1240 The alchemist Arnold of Villanova (Spain) discovers carbon monoxide when he notices that burning wood without adequate ventilation produces poisonous fumes. Also makes the first preparation of pure alcohol.

1267 One of the earliest advocates of the scientific method, Englishman Roger Bacon writes that it is wrong to rely only on the authority of past scholars and that experimentation is an important means of gaining knowledge. Performs experiments on optics, recognizes the visual spectrum of light and manufactures explosives. Imprisoned for 10 years on charges of witchcraft.

1086 Chinese scientist Shen Kua's "Dream Pool Essays" outline the first principles of earth science: erosion, uplift, and sedimentation. Also contain a detailed description of how to make a magnetic compass. Chinese navigators begin to use compasses in the following decades.

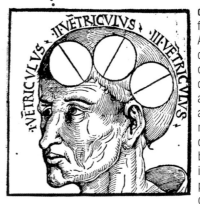

c. 1250 German friar and scholar Albertus Magnus describes his observations and dissections of many animals and insects and ridicules the mythical animals described in bestiaries. Also investigates vision phenomena such as colorblindness, and contemplates brain function and cognition.

1000

1269 Considered the first experimental physicist, Petrus Peregrinus writes of his experiments on magnetism, involving the forces between a magnet and a compass dial (like one pictured from 1570). His writings contain the first description of the polarity of magnets.

1316 Mondino de Luzzi (aka Mundinus) publishes a book about human anatomy (title page, pictured) and the art of dissection, two topics abandoned for centuries in European medical training. Performs dissections at public lectures. Although his book contains some inaccuracies, it's the standard textbook for 300 years.

Science Clashes with Religion

In 1210 the teaching of Aristotle's work is forbidden at the University of Paris as a pagan threat to Christianity. In the 1250s St. Thomas Aquinas tries to reconcile Aristotle and Christianity, reason and religion. But in 1277 the Bishop of Paris condemns as heresy many of Aristotle's ideas, which explain the natural world without relying on a divine power. Tensions between religion and science continue and in 1317, Pope John XXII issues a prohibition against alchemy.

1284 The Italian Salvino D'Armate is credited with inventing wearable eyeglasses, an improvement on the polished glass blocks previously used as reading aids. Convex lenses correct for farsightedness—glasses that correct for nearsightedness aren't invented until the 15th century.

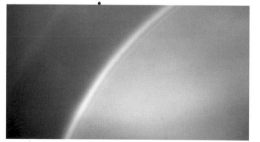

1304 The Dominican monk Theodoric of Freiburg (Germany) publishes "On the Rainbow," which explains rainbow formation based on experiments with globes of water. Says that rays of light are refracted, or bent, as they pass from air to water.

1350 French philosopher Jean Buridan develops the idea of impetus, rejecting Aristotle's idea that a projectile is propelled by vibrations in the air. Buridan says the thrower gives an object impetus, which carries it forward until the impetus is diminished. Anticipates Galileo's theory of inertia: that a body remains at rest unless a force compels it to move.

c. 1400 Oil-based paints are developed, paving the way for the artistic masterworks of the Renaissance. Painter Jan Van Eyck in Holland helps bring them to prominence in the 1420s. Painters boil the seed of the flax plant to make linseed oil, which they mix with pigments.

1435 Leone Battista Alberti of Genoa (Italy) publishes a book on drawing that discusses the laws of perspective and basic geometry applications. Influences Renaissance artists struggling to give their paintings the illusion of three-dimensional reality, like in this famous painting, "The School of Athens," by Raphael.

1400

1377 First quarantine station set up in the port of Ragusa (Yugoslavia) in the midst of the black plague. Newcomers to the city spend 30 days in an isolated location, where they are watched for plague symptoms. Isolation period is later lengthened—the word quarantine comes from the Italian "quaranta giorni," meaning 40 days.

1464 German astronomer and mathematician Regiomontanus writes a book advancing the understanding of trigonometry, the study of triangles. Applications of trigonometry include triangulation, which is used in navigation, astronomy, and optics.

The Book Boom

Johann Gutenberg prints the Bible in Germany in 1455 using a new system of move-able type, a vast improvement on the laborious block printing method. This event ushers in an era of cheaper book production and more accessible information, which affects every discipline of learning. (Chinese invented moveable type earlier, but it was not in widespread use. Gutenberg is thought to have invented his technique independently.)

1480–1500

In his notebooks, Leonardo Da Vinci describes and draws intricate details of human anatomy. Also sketches many ideas for inventions, including a workable hang glider, a helicopter, an armored tank, a clock with a pendulum, and a submarine. Most of his ideas are not acted upon, and the devices are not built.

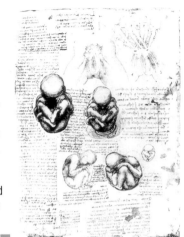

1500

Jakob Nufer of Switzerland performs the first recorded caesarean operation on a living woman. Nufer, a pig gelder, operates on his wife after a prolonged labor. Both she and the baby survive.

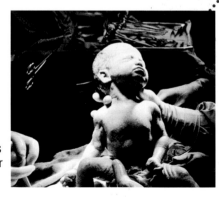

1507 In Germany, the first map is published calling the continent Columbus discovered "America." Named after the merchant and cartographer Amerigo Vespucci who recognizes it as a new continent and not part of Asia.

1514 Nicolaus Copernicus writes the first version of his heliocentric theory of the solar system, although it's not published until 1543, just before his death. Recognizes that the earth and other planets orbit the sun, and the moon orbits the earth. Discards Ptolemy's theories, as they don't agree with observations of the movements of the planets.

From Laudanum to Darwin: 1515 to 1859 A.D.

In earlier eras, bursts of scientific advancement occurred across the globe, with South American, Asian, and Middle Eastern societies pushing forward by turns. But by the 1500s, the European nations dominated scientific pursuits.

The early Renaissance in Europe began during the 14th century and was defined by the rediscovery of the cultures of ancient Greece and Rome. European scholars revived philosophies, scientific theories and artistic techniques from over one thousand years before. Initially, scholars embraced the wisdom of the ancients without testing its accuracy, setting up a reaction against accepted doctrine in the later Renaissance. Scientists continued to revere Aristotle as a philosopher, but began to question his theories on the composition of matter and the physical laws of motion.

The work of the Flemish anatomist Andreas Vesalius in the 1530s demonstrated the new mindset. As a professor at the University of Padua in Italy, he rejected the previous method of teaching surgery, in which students read the classic texts by Galen, a physician from the 2nd century. Vesalius pointed out that human dissections had been banned in ancient Rome and that Galen had instead dissected Barbary apes, leading to numerous errors in his assumptions about human anatomy. In Vesalius's classroom, the students clustered around the dissection table while their professor gave them a tour of the human body. It was a triumph of empirical evidence over conventional wisdom.

In retrospect, we call the era beginning in the early 1500s the Scientific Revolution, when scholars stopped looking for answers through logical deduction and turned instead to experimentation and the evidence of their senses. Sir Francis Bacon, an English philosopher, is not famous for conducting breakthrough experiments of his own, but rather for popularizing what would come to be called the scientific method. "If a man will begin with certainties, he shall end in doubts; but if he will be content to begin with doubts he shall end in certainties," he wrote in 1605. His call for scientists to test their hypotheses by experimentation and observation resonated through the fields of astronomy, physics, and the life sciences.

Outside of Europe, naval exploration was encouraged for a time in China during the Ming dynasty. In the early 15th century, the legendary explorer Zheng He led seven expeditions of his "Treasure Fleet" to present-day Vietnam, India, and down the east coast of Africa. After Zheng He's death in 1433, a new emperor decommissioned the fleet and banned new shipbuilding, deciding to concentrate instead on domestic affairs. Scholars have argued that this insularity gave way to distrust of new ideas that held back scientific progress in China for many centuries.

The Ottoman Empire was the other great power of the era, centered on the city of Constantinople (now Istanbul in Turkey). It reached its zenith under the sultan Suleiman the Magnificent in the mid-16th century, when its borders stretched from the northern coast of Africa to the Black Sea, and the cities thronged with artists, craftsmen, and poets. However, the sultans placed more value on commerce and culture than on science, and the Empire lagged behind Europe in scientific knowledge.

As the expanding Ottoman Empire blocked overland routes used by merchants to reach the Far East, European traders were forced to take to the sea. The Spanish and Portuguese excelled at shipbuilding and navigation, and began launching bold men with new theories on how to reach the Spice Islands (Indonesia) by sea. Christopher Columbus crossed the Atlantic in 1492; 30 years later, in 1522, the bedraggled remains of Ferdinand Magellan's expedition returned from the first successful voyage around the globe.

The "Age of Exploration," which continued into the 17th century, shifted the balance of power in Europe. Those countries that dominated maritime exploration, namely Spain, Portugal, Britain, and France, grew wealthy from the resources extracted from their colonies. In these countries, the pace of scientific discoveries and accomplishments quickened, pushed onward by a new sense of possibility. If sailors could find unknown continents beyond the blue horizon, who knew what scientists could discover? Scholars pointed improved telescopes at the dark sky and slipped slides of pond water underneath the magnifying lenses of new microscopes. They were amazed by what they saw.

The new tools of science brought a flood of novel observations, and the work of scientists like Sir Isaac Newton brought a new rigor to the analysis of these results. Newton's *Principia*, published in 1687, not only contained many breakthroughs in physics, but also called on scholars to quantify the results of their experiments, and to analyze those results mathematically. The calculus that Newton and the mathematician Gottfried Leibniz independently developed allowed researchers to write formulas to express an object's motion, the movement of energy, heat, and light, and other natural phenomena.

In the mid-1600s, groups of scholars founded the first scientific societies in Italy, France, and England. Originally conceived as informal gatherings to exchange ideas, the groups soon became more formally organized and began sponsoring regular lectures and publishing scientific journals. The Royal Society, which was organized officially in 1660, began publishing *Philosophical Transactions* in 1665; previously scientists kept abreast of developments in their field and announced their discoveries in personal letters. By the 1700s, copies of the journal were mailed to interested scholars in foreign lands, leading to a wider spread of ideas and the birth of truly international science.

The societies' lectures served another important function—spreading scientific ideas to the wider public. The most famous debate over Charles Darwin's theory of evolution occurred at a meeting of the British Association for the Advancement of Science in Oxford. Darwin's 1859 book, *On the Origin of Species by Means of Natural Selection*, stirred up a heated public controversy with its discussion of evolution and its implication that God didn't create human beings in their present form. At the 1860 debate, the Bishop of Oxford arose and mockingly asked a Darwin defender whether he was descended from apes on his grandmother's or grandfather's side. The defender, the biologist Thomas Huxley, allegedly replied: "I would rather be descended from an ape than from a cultivated man who used his gifts of culture and eloquence in the service of prejudice and falsehood."

1520 Swiss physician and alchemist Philippus Aureolus Paracelsus introduces an alcoholic tincture of opium to medicine. He calls it laudanum and suggests its use as a painkiller, unaware of its addictive properties.

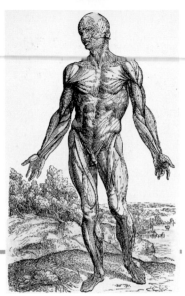

1537 Italian mathematician known as Niccoló Tartaglia (translated as Niccolo the Stutterer) publishes a text about ballistics, the study of firearms and projectiles. States Tartaglia's theorem: the trajectory of a projectile is a curved line, and by firing a projectile at an elevation of 45 degrees, it will travel maximum distance. His hometown of Verona quickly applies his findings in its military strategies.

1543 Belgian anatomist Andreas Vesalius's seven-volume text, "De Humani Corporis Fabrica," revolutionizes the understanding of the human body. The most comprehensive view of the human anatomy of its time, greatly surpassing the teachings of the highly regarded anatomist Galen (2nd cent. A.D.), whose work was based on animal dissections. Vesalius shows that the heart has four chambers and that the blood vessels originate in the heart, not the liver.

The Roman Catholic Inquisition

In 1542, Pope Paul III establishes the Holy Office to defend the faith and condemn heresies. The vigilant group censures mystics, clerics, and scientists, most famously finding fault with Copernicus's heliocentric theory of the solar system; in 1616, the Holy Office calls the idea that the sun doesn't move around the earth both "foolish and absurd" and "formally heretical."

1570 Giambattista della Porta experiments with the camera obscura, or pinhole camera: when light passes through a small hole in a wall, it projects an upside-down image of the outside world onto the opposite wall. In his series of books on popular science, "Natural Magic," della Porta explains how to put a convex lens inside the aperture to brighten the projected image, spreading the device's popularity.

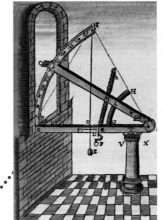

1577 Brahe tries to determine the distance of a comet from earth from his observatory on the island of Hven between Denmark and Sweden. His crude observations show the comet is at least four times as distant as the moon, proving that comets aren't an atmospheric phenomenon.

The Scientific Revolution

The European Renaissance, with its embrace of learning and creativity, gives rise to the Scientific Revolution around the end of the 16th century. Scientists begin conducting experiments to examine theories of the natural world proposed by ancient scholars, and often disprove those cherished theories. They soon begin to place more faith in their own observations than in the inherited wisdom of the past, and a growing thirst for knowledge drives researchers to examine unexplained phenomena such as magnetism and electricity.

1572 A supernova—an exploding star—appears in the constellation Cassiopeia. Danish astronomer Tycho Brahe observes it, and uses it to challenge the prevailing view of stars as unchanging and eternal. Danish King Frederick II is impressed with Brahe's work, and funds two new observatories for him on an island off the Danish coast.

c. 1595 Dutch spectacle-maker Hans and his son Zacherais Janssen are thought to have invented the compound microscope around this time. They combine two double convex lenses in a tube to increase magnifying power.

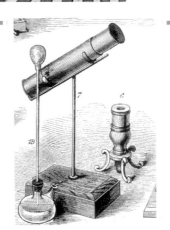

1597 Francis Bacon publishes essays describing the process of inductive reasoning, in which observations and experiments are studied to arrive at the general laws of nature. A rebuttal to the Aristotelian system of deduction, wherein established "known facts" are analyzed for further understanding. Widely interpreted as the beginning of formal scientific thought in Europe.

1603 Italian physician Girolamo Fabrici, better known as Hieronymus Fabricius, discovers that leg veins have valves that allow blood to flow only toward the heart. Later shows that blood circulates to the fetus through the umbilical cord.

1600

1600 English physician William Gilbert shows that magnetism is an earthly phenomenon, rather than one governed by the heavens, in a text called "On the Magnet." One of the first true experimenters, Gilbert relies on observation to debunk misconceptions such as the folk belief that garlic destroys the magnetic property of a compass. Also the first to distinguish between magnetism and static electricity.

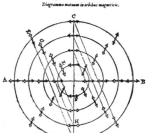

1609 Galileo hears of a telescope recently invented by a Dutch scientist and builds his own, with a magnifying power of 20. Over the ensuing year he observes the moon's mountains, sunspots, the rotation of the sun, and shows that the planets are closer to earth than stars. Also recognizes the Milky Way as a collection of faint stars.

Galileo Galilei

One of the brightest lights of the scientific revolution is Galileo Galilei, a professor of astronomy, mechanics, and geometry at the University of Padua. He is one of the first astronomers to point a telescope toward the stars, and makes numerous discoveries. He also sets a new, rigorous standard for experimental physics, insisting on results that can be quantified and analyzed mathematically in his experiments on falling objects and pendulum swings. Finally, he advances the philosophy of science by encouraging researchers to disregard doctrine—whether it comes from the church or from revered scholars of the past—and to trust the evidence of their own senses.

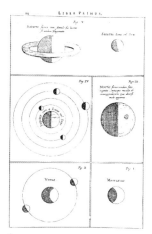

1609 In his revolutionary text, "New Astronomy," Johannes Kepler shows that the planets have elliptical orbits. A former assistant to Tycho Brahe, Kepler uses Brahe's very accurate observations of planetary positions to arrive at his laws of planetary motion. Challenges the common belief, proposed by Ptolemy, that the orbits of planets are perfect circles.

1620 Johann Baptista van Helmont coins the term "gas" (possibly as a corruption of "chaos") to describe air-like substances. Recognizes that gas produced by burning charcoal is the same as that produced in fermentation: he calls it gas sylvestre, but it's later named carbon dioxide.

The 17th and early 18th centuries bring a burst of scientific inventions, including telescopes, improved microscopes, pendulum clocks, thermometers, and barometers. The new magnifying lenses in telescopes and microscopes let scientists make more detailed observations. Experimenters also begin recording phenomena in terms of values measured (such as seconds, temperature, and pressure), and gain the ability to compare results precisely.

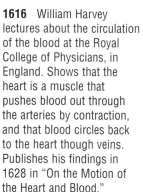
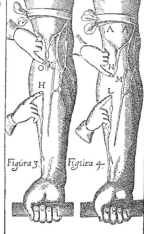

1616 William Harvey lectures about the circulation of the blood at the Royal College of Physicians, in England. Shows that the heart is a muscle that pushes blood out through the arteries by contraction, and that blood circles back to the heart though veins. Publishes his findings in 1628 in "On the Motion of the Heart and Blood."

1623 German mathematician Wilhelm Schickard invents a calculating machine before contemporaries such as Liebniz and Pascal (whose version is pictured). The machine, which can add and subtract 6-digit numbers, is later used by Johannes Kepler to calculate astronomical tables.

1633 Roman Catholic Inquisition summons Galileo to Rome for a second time, and forces him to recant his Copernican view that the Earth moves around the sun, which had formerly been declared heretical. The authorities ban his 1632 book that supports the Copernican system, and Galileo spends the rest of his life under house arrest. The Church won't remove Galileo's work from the prohibited list until 1822.

Beyond Aristotle

By the 1600s, many of Aristotle's theories about the natural world had outlived their usefulness, and bold thinkers began to question the assertions of the revered Greek scholar. In physics, Aristotle had said that the speed of falling bodies is determined by the proportions of the four prime elements they contain (earth, water, air and fire). Galileo disproves this idea, and states that in a vacuum, all bodies would fall at the same speed, regardless of their mass or composition.

1637 The French philosopher René Descartes (portrait of him by Frans Hals) publishes the book *Discourse on Method* to explain his "rationalist" philosophy. Develops a philosophical framework for the natural sciences, and advises scientists on how to arrive at objective truths about the world. In his description of analytic geometry, an important precursor to calculus, he says, "Nature can be defined through numbers."

1643 A student of Galileo's, Italian physicist Evangelista Torricelli, invents the barometer to measure changing air pressure in the atmosphere. A few years later, French mathematician Blaise Pascal famously proves the functionality of barometers by having his brother-in-law carry one up a mountain and record its measurements, showing that higher altitudes have lower air pressure.

1650 German physicist Otto von Guericke invents an air pump he uses to produce a vacuum. Later demonstrates the strength of a sphere held together by a vacuum with two teams of horses (16 total), which are unable to pull apart the two halves of the sphere.

1656 Dutch astronomer Christiaan Huygens observes Saturn's rings. First described as "ears" by Galileo, who was puzzled by their disappearance after several years, Huygens shows the rings are positioned at an inclined angle around the planet, and appear to vanish when oriented directly at the earth. Also discovers Saturn's largest moon, Titan.

1658 Dutch microscopist Jan Swammerdam is thought to be the first scientist to observe and describe red blood cells: he sees oval shapes in a sample of frog's blood. While his research in anatomy continues, he is later known mostly for his work classifying insects and studying their metamorphoses.

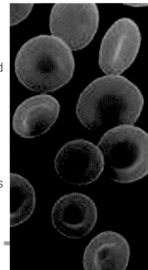

1661 Robert Boyle publishes "The Sceptical Chymist," which makes progress toward analyzing the components of chemical compounds, and advances the idea of elements that can't be broken down further. Also rejects the Aristotelian idea that everything is composed of only four elements: earth, water, air and fire.

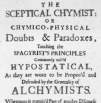

1656 In a breakthrough in timekeeping, Christiaan Huygens patents the pendulum clock based on experiments done by Galileo. First clock accurate enough for scientific use—before this, experimenters used their pulses or inaccurate water clocks to measure time.

1660 Supported by King Charles II, a group of scholars founds the Royal Society (their coat-of-arms, pictured) to facilitate discussion of scientific theories and experiments. The Society, and others like it in France and Italy, encourages the exchange of ideas and formalizes the peer review process, where colleagues check each other's work. The Royal Society is still in existence today.

31

Alchemy's End

Although Robert Boyle still considered himself an alchemist and made many attempts to transform common metals into gold, his chemical research was instrumental in removing the superstitious underpinnings of chemistry. His work is continued in the 18th century by chemists such as Antoine Lavoisier, who eventually shows that elements can't be transformed into other elements, but only combined to form chemical compounds.

1667 Duchess of Newcastle, Margaret Cavendish, insists on joining the Royal Society, England's scientific society. Becomes known as Mad Madge for her fascination with science and eccentric ways. In one of her many books she draws a distinction between wit and learning; says that women have as much wit (or natural intelligence) as men, but are denied learning. The only woman admitted to the Royal Society until 1945.

1669 In his study on fossils and geology, the Danish physician Nicolaus Steno recognizes that fossilized seashells and shark teeth have a biological origin. To explain why many fossils are found far inland in rocks above sea level, he theorizes that layers of sediment were laid down on the seafloor over time, and were subsequently lifted and tilted by the earth's upheavals.

1660

1665 Robert Hooke, an English biologist, publishes his book "Micrographia" which contains both microscopic and telescopic observations. Coins the term "cell" because the lines of tiny plant cells, viewed in a slice of cork, remind him of the cells or small rooms where monks sleep. The book is an immediate best seller, and sparks public interest in the potential of microscopy.

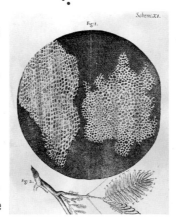

1668 Italian scientist Francesco Redi disproves the prevailing idea that maggots arise spontaneously from rotten meat. Does so through experiment that prevents flies from landing on the meat to lay their eggs. One of the first controlled experiments in history, leads to the demise of the dominant theory of "spontaneous generation." Ironically, Redi still believed that spontaneous generation explained the existence of some specimens, such as intestinal worms.

1669 Naturalist John Ray, sometimes called the father of English natural history, publishes results of his experiments on trees, showing that sap ascends through the wood. His many works on botany show the diversity of plants in England and Europe, and advance the method of classifying plants according to their similarities.

1679 German mathematician, Gottfried Leibniz, introduces binary arithmetic by showing that every number can be represented by combining the symbols 0 and 1. Does not publish his findings until 1701, when he sends his work, "Essay on a New Science of Numbers," to the Paris Academy to celebrate his induction to the Academy. This binary system is used today in virtually all computers.

01001
10110
10100
00001

1672 The great astronomer Giovanni Cassini, with the aid of his colleague, Jean Richer, calculates the distance from Earth to Mars. In Paris, Cassini observes the position of Mars against distant stars while Richer made simultaneous observations in French Guyana; they then use trigonometry to work out Mars's distance from Earth. Cassini goes on to determine the distances of all the planets from the sun, and thus, the dimensions of the solar system.

1675 Ole Römer, a Danish astronomer, measures the speed of light by studying the eclipses of Jupiter's moons at different times of the year—when the Earth is farther from Jupiter, the moons appear to take longer to orbit Jupiter. Römer concludes that the difference is due to the extra distance the light has to travel to reach Earth. He estimates the speed of light at 140,000 miles/second; it's later determined to be 186,292 miles/second.

33

1683 The first sighting of bacteria by Dutch amateur scientist Anton van Leeuwenhoek, who uses a homemade microscope with the unprecedented magnifying power of 250. Also discovers single-celled organisms (protozoa) in a sample of pond water, which he calls "cavorting beasties," and confirms the discovery of sperm, showing that they're not evidence of disease or of the putrefaction of semen.

1687 In Isaac Newton's masterwork, "Mathematical Principles," the English mathematician publishes his law of universal gravitation: states that the same attraction that causes apples to fall to the ground keeps the moon in orbit around the earth, and the planets in orbit around the sun. To express the motions of the planets in mathematical terms, he develops calculus (Gottfried Leibniz independently arrives at another version).

Calculus: The Language of Physics

The development of calculus by Newton and Leibniz helps scientists discuss processes that had previously been poorly understood. It begins with Newton's description of motion as a "rate of change." Physicists soon realize other phenomena can be described using similar formulas, including heat, light, electricity, and magnetism.

1686 French writer Bernard Fontenelle popularizes the ideas of astronomers, from Copernicus to Descartes, in his famous text, "Conversations on the Plurality of Worlds." States that the sun is but one of a myriad of stars.

"Our world is terrifying in its insignificance," he writes.

1694 German professor Rudolph Camerarius publishes his findings that plants reproduce sexually, and that pollen from male plants must be deposited on the stigmas of female plants in order to produce seeds. Understanding this process paves the way for deliberate cross-pollination and species hybridization.

1698 Denis Papin, a former assistant to Christiaan Huygens, builds the first steam engine, based on work he published in 1690. Later uses his engine to raise water in a canal, and to pump water into a tank on the roof of a palace to supply its fountains. Soon after, the "Miner's Friend" is invented in England—powered by steam, it pumps water from coal mines.

The Newtonian Era

The Age of Enlightenment dominated European thought in the 18th century, characterized by a rational approach to political, religious and scientific concerns. In the scientific realm, Isaac Newton's work defined the era. He advanced the fields of physics, astronomy, and mathematics, but his approach to science has as much influence as his theories. In his scientific method, he insists on only proposing hypotheses that can be proven through observation and experimentation, rather than putting forward grand, unverifiable theories about the universe and its forces.

1700

1704 Isaac Newton's "Opticks" declares that white light passed through a prism produces a spectrum of colors, which are its component parts. Previously, Roger Bacon had stated that the prism added the colors. Newton also states that light is particulate in form; two centuries later scientists understand that light has properties of both waves and particles.

1705 After ten years of study, English astronomer Edmund Halley correctly predicts that a comet that last appeared in 1682 will return in 1758—it comes to be known as Halley's Comet. His proposal is based on theory that comets have elliptical orbits, which clashes with Newton's belief that comets had parabolic paths that bring them through our solar system only once.

1714 German physicist Gabriel Fahrenheit invents a mercury thermometer with a temperature scale that will be named after him. (a later version, from c. 1790, pictured.) Originally uses alcohol in his thermometers, until he discovers that mercury has a more constant rate of expansion when heated, thus providing more accurate measurements.

1717 Italian anatomist Giovanni Lancisi, often considered the first major hygienist, suggests that malaria is transmitted by mosquitoes in his book "On the Noxious Effluvia of Marshes." He advises draining marshes and swamps near cities as a solution to malaria outbreaks.

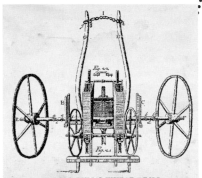

JETHRO TULL'S WHEAT DRILL

1728 Pierre Fauchard publishes "The Surgeon Dentist" in France, the first scientific treatise on dentistry. Declares that not all teeth with cavities need be extracted, and describes how to fill a cavity with tin, lead, or gold. Rebuts the theory that cavities are caused by "tooth worms," and advises people to limit their consumption of sugar for healthy teeth.

Electricity

In the 17th century, scholars begin experimenting with static electricity, which is viewed as a strange curiosity. In the early 18th century, the self-taught scientist Stephen Gray discovers that what he calls "electric virtue" can be conducted along a long thread to an ivory ball, which will then attract light objects through static electricity. A few years later, a device called the Leiden jar is invented for storing static electricity, and scientists find that anyone who touches the metal conductor emerging from the jar releases the electricity and gets a nasty shock.

1733 Jethro Tull's "Horse-Hoeing Husbandry" advocates using manure in the fields, pulverizing the soil, growing crops in rows, and hoeing to remove weeds, all of which increase crop yields. Also invents a seed drill to speed up planting and increase the percentage of seeds that germinate and take root. All part of the British agricultural revolution.

1736 At the new University of Göttingen, Swiss physician Albrecht von Haller founds an anatomical theater and an obstetrical school, which contributes to the medical understanding of childbirth and its complications, a field often left to midwives. Other contributions include better understanding of birth defects, and a demonstration of how nerve impulses cause muscles to contract.

Poma Aurantia nana dicta.

Mala Medua seu Citria.

Poma Aurantia.

1747 The Scottish physician James Lind demonstrates that scurvy, a disease common to sailors, can be cured by eating citrus fruits. The first "dietary deficiency disease" recognized, although vitamin C and the role it plays in good health aren't fully understood until the 1930s.

1735 In his book, "Systems of Nature," Swedish biologist Carolus Linnaeus introduces the system for classifying organisms still in use today. The first to group organisms into a classifying genus based on shared biological similarities—like body shape and size—and then assign a specific species name to each distinct type of organism. Later describes over 10,000 species in his books.

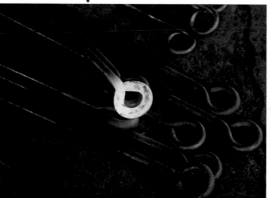

1744 Russian chemist Mikhail Lomonosov publishes a paper in Russia on the causes of heat and cold. He correctly posits that heat is a form of motion—the transfer of thermal energy—a step toward the systematic study of thermodynamics. Also anticipates the law of conservation of mass, which states that matter changes form, but can't be created or destroyed.

Benjamin Franklin

Benjamin Franklin is not only a leading figure in the American Revolution and an influential diplomat in England and France, he is also America's premier scientist and inventor. His experiments on electricity are copied throughout the world, and his inventions include bifocal glasses, the medical catheter, and the Franklin stove, a woodstove which generates heat more efficiently than the fireplaces in common use. Because he believes in the free exchange of ideas, Franklin never patents any of his inventions.

1754 The University of Halle in Germany graduates Dorothea Erxleben, the first woman with a degree as a medical doctor. Since universities don't admit women, she petitioned the Prussian king to allow her admission to the school. Alarmed male doctors accuse her of witchcraft, and suggest it should be illegal for women to be doctors.

1769 Englishman Richard Arkwright invents a machine that spins wool. It's too large and expensive for use in cottages, and in 1771 he builds the first factory, a water-powered cotton mill, to house it. One of the first events in the Industrial Revolution, in which production shifts from small batches of goods made in homes or workshops to the mass production of goods made in factories.

ARKWRIGHT'S SPINNING-FRAME.

1749 Founding Father, scientist and inventor Benjamin Franklin installs a lightning rod on his house in Philadelphia, showing that the rod can serve as a conductor that draws off an electric charge, thereby protecting his home from a lightning strike. A few years later, his famous kite experiment proves that lightning is a form of electricity.

1766 English physicist and chemist Henry Cavendish recognizes that hydrogen is a discreet element, which he calls "inflammable air." Later named by French chemist Antoine Lavoisier when he repeats Cavendish's experiments and proves that water (H_2O) is composed of hydrogen and oxygen.

Georg Cuvier bei der Bestimmung fossiler Tierreste
Nach dem Gemälde von Chartran

1780 The huge skull of a creature later called a "Mosasaur" is found in a stone quarry in the Netherlands. The first remains of a giant, prehistoric reptile to be so identified when examined by Georges Cuvier in 1795.

ASCENT OF M. GIBARD'S MONTGOLFIER BALLOON FROM CREMORNE GARDENS.

1783 Brothers Joseph-Michael and Jacques-Etienne Montgolfier send two men aloft in their hot air balloon. Made of taffeta and coated with varnish, it soars over Paris for 20 minutes—man's first flight.

The Nature of Fire

Antoine Lavoisier's studies of combustion lay to rest the theory of phlogiston; earlier chemists believed combustible materials contained this invisible substance, which was released on burning. In his precise and careful experiments, Lavoisier shows that phosphorus and sulfur absorb air during combustion instead of expelling something. He does this by weighing the elements before and after combustion, showing that they're heavier after the reaction has taken place. Lavoisier both dispels a long-held misconception and sets a new standard for chemical experiments.

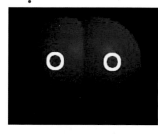

1772 Carl Scheele in Sweden and Joseph Priestley in England independently identify the element oxygen, called "fire air" by Scheele. Priestley notes that breathing pure oxygen makes respiration "peculiarly light and easy." Antoine Lavoisier later names it oxygen and determines that combustion is a chemical process that occurs when oxygen combines with a substance, oxidizing the substance and releasing heat.

1781 William Herschel, an amateur astronomer, observes what he initially believes to be a comet. Realizing that the object's movement is too slow and steady and its orbit too circular to be a comet, he identifies the object as the distant planet Uranus. Discovery doubles the size of the known solar system.

The Industrial Revolution

While it's difficult to say when the Industrial Revolution begins, by the end of the 18th century it is clearly underway in Europe. The first spinning machines are in use, the first factories are being built, and steam engines are used to pump water out of mines. Early industrialists, excited by the gains in efficiency and productivity brought about by these new technologies, encourage further research and accelerate the societal change. People begin to leave the countryside and flock to the cities for factory jobs.

1789 Antoine Lavoisier publishes a table of 31 chemical elements, simple substances that can't be broken down any further. In addition to elements such as hydrogen, nitrogen, oxygen, and mercury, he wrongly includes heat and light as elements. Also published drawings of equipment used in his experiments, like apparatus used to investigate the chemical composition of water (pictured).

1796 English physician Edward Jenner inoculates an 8-year-old boy with cowpox, a mild disease common in cattle, and six months later shows the boy can't catch smallpox, a far more dangerous human disease. His vaccination method, named by Jenner after the Latin word for cow, "vacca," is quickly adopted throughout Europe, greatly reducing the death toll of disease outbreaks. In the early 19th century, governments begin passing laws mandating vaccination for all citizens.

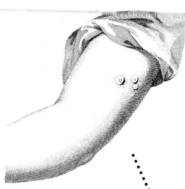

1791 While dissecting a frog, Italian physician Luigi Galvani touches an exposed nerve with his scalpel, which had picked up an electric charge, and sees the frog's leg kick. First step to understanding bioelectricity: the voltage of biological cells and the electric currents that flow through nerves and muscles.

1794 After being found guilty of treason during the Reign of Terror in post-revolutionary France, pioneering French chemist Antoine Lavoisier is beheaded. When he asks the presiding judge for more time to finish his experiments, the judge allegedly responds, "The Republic has no need for scientists."

1797 Caroline Herschel, a German astronomer living in England, discovers her eighth comet in 11 years. As an assistant to her brother, who is an astronomer for King George III, she receives an annual salary from the king, which makes her the first woman officially recognized as a professional scientist.

Caroline Herschel

1799 Italian physicist Alessandro Volta invents the first electric battery, called the Voltaic pile. The stack of alternating silver and zinc disks separated by felt soaked in brine is the first source of steady electric current. It's used primarily in experiments with electricity and with electrolysis, when a compound is broken down into its component elements using an electric current.

Fig. 255. Pile à colonne construite par Volta en 1800.

1800

1798 Cavendish calculates one of the first accurate estimates of the Earth's mass. Using Newton's formula for determining gravitational force, Cavendish plugs in the values for an object's mass and the gravitational attraction between the object and the Earth, and then solves the equation for the mass of the planet. Cavendish's calculation remains the standard until well into the 20th century, and differs only slightly from the current mass estimate of 5.98 billion trillion tons.

1801 Thomas Young demonstrates the interference phenomena of light: light passing through two narrow, parallel slits in an opaque screen produces a pattern of bright and dark lines, where the light waves emerging from the two slits interfere with each other. Seen as proof that light is a wave phenomenon. A century later, scientists understand that light has properties of both waves and particles.

1804 Nicholas de Saussure publishes a book in France challenging the widespread belief that plants get the carbon vital to their growth from the soil. He shows that growing plants collect carbon from carbon dioxide in the air, and take water and nitrogen from the soil.

1807 Robert Fulton's steamboat, the *Clermont*, makes its maiden voyage from Manhattan to Albany up the Hudson River. The trip takes thirty-two hours, with the vessel averaging about five miles an hour. Though Fulton did not invent the first steamboat, his is the first practical and economic one.

1808 English chemist John Dalton publishes his universal atomic theory in the first volume of his work, "New System of Chemical Philosophy." Proposes that all matter is composed of very small particles called atoms, that each element has its own distinct kind of atom, and that atoms combine in definite proportions to form molecules.

Early Theories of Evolution

Charles Darwin was not the first to believe that the physical characteristics of animals change over time. An earlier, vocal proponent of evolution was the French biologist Jean-Baptiste de Lamarck, but he got the mechanism wrong. He thought individual animals acquired characteristics based on need—that giraffes grew longer necks to reach the leaves in the tree-tops, for example—and then passed those traits down to their descendants.

1807 London's streets begin to be illuminated by a coal-gas lighting system invented by William Murdoch, a Scottish engineer and inventor. Gas lamps, which are considerably cheaper than oil lamps or candles, soon become widespread, and their availability encourages reading and scholarship among the public. (Pictured: gaslights on the streets of New York City.)

1812 Benjamin Rush, a physician and founding father of the United States, publishes "Medical Inquiries and Observations upon the Diseases of the Mind." One of the first attempts to explain mental illness as a physical disease rather than a supernatural possession. He describes conditions such as hypochondria, manic depression, and "tristimania," which can be compared to clinical depression; his recommended cures include bloodletting, distraction, and alcohol.

1814 At the Killingworth coal mine, British engineer George Stephenson introduces his first steam locomotive. It's capable of hauling 30 tons, moves at 4 miles per hour, and has flanged wheels that run on top of the iron rails. Steam-powered trains quickly improve in efficiency, and are commonly used to transport passengers and freight by the 1830s.

1820 Danish physicist Hans Christian Oersted accidentally discovers that a magnetized compass needle is deflected from pointing north when the electrical current in a nearby battery is turned on. Subsequent experiments show that an electric current produces a magnetic field as it flows through a wire. The beginning of the study of electromagnetism.

1825 Pierre Bretonneau successfully performs the first tracheotomy—the surgical incision of the trachea through the neck—to restore breathing to a child suffering from diphtheria. He began to study the disease, wherein a membrane forms in the throat and obstructs the airway, after Napoleon Bonaparte loses a nephew to the disease and offers a prize for research.

1815 William Smith publishes the first comprehensive geological map of England and Wales, showing features such as rock strata, mineral deposits, and mines. Smith conducts his research while working as a surveyor in coalmines and a canal engineer. He observes that rock layers are arranged in the same patterns across wide geographical areas, and realizes that particular strata can be identified by the fossils they contain.

1822 Joseph Nicéphore Niépce produces the first fixed positive image that could be called a photograph, using a pewter plate covered with a petroleum derivative. The slow process requires about 8 hours of bright sunshine to fix the image, therefore Niépce takes pictures of buildings and inanimate objects. Better, faster processes soon replace his technique.

1825 French physiologist Marie Jean Pierre Flourens studies the nervous system and the brain by removing parts of the brains of living rabbits and pigeons, and observing the effects on sensibility and behavior. Demonstrates that the cerebellum controls muscular movement, and that the brainstem controls vital functions such as the heartbeat and respiration.

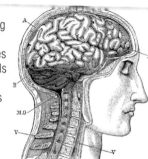

1828 Friedrich Wöhler, a German chemist, synthesizes the organic compound urea by mixing the chemicals ammonia and cyanic acid, proving wrong the idea that organic compounds can be produced only by living organisms. Previously, scientists thought a "vital force" animated organic compounds.

1831 While studying orchids, Scottish botanist Robert Brown notices a dark structure within the plant cells, which he names the nucleus, meaning "little nut." Microscope users had sighted this feature before, but Brown was the first to observe the nucleus in the cells of many different plant species, and to call it a component of the cell.

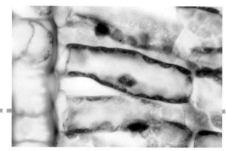

1827 Russian biologist Karl Ernst von Baer reports his discovery of eggs in mammals, and states that embryos develop from these ova. While studying amphibians, he also discovers the blastula stage of their embryos, when the fertilized ovum divides into a mass of undifferentiated cells.

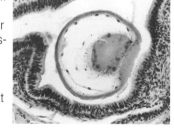

1830

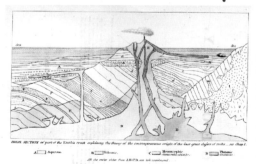

1830 The first volume of *The Principles of Geology* (frontispiece shown) by Charles Lyell (Scottish) begins a massive study showing that Earth must be several hundred million years old. Spreads ideas about coal formation, glaciers, and volcanic activity, and advances the theory of uniformitarianism: that the earth was shaped by gradual forces acting over long periods of time, which are still ongoing. Rebuts the "catastrophic" theory that the earth's features were caused by periodic cataclysmic events, such as massive earthquakes and floods.

1831 American chemist and physicist Samuel Guthrie discovers chloroform, the first practical anesthetic, and develops a process for producing it. Not commonly used until the late 1840s, when doctors use it in surgeries and childbirth. Abandoned in the early 20th century due to numerous side effects.

1835 American physicist Joseph Henry invents the electrical relay, enabling a current to travel long distances from its origin. He previously developed a powerful electromagnet that has strong magnetic force even when it receives electricity from a small battery through a mile of wire. Then demonstrates to his students at Princeton University that a series of electromagnets can be strung together with wire.

The Nature of Heat

In the middle of the 19th century, the science of thermodynamics found its scientific footing. The reigning theory until then stated that changes in temperature were due to the transfer of an invisible and weightless fluid called caloric. Experiments soon rebut this idea with the "kinetic theory" that heat is a form of motion. In 1842, German physician Julius Mayer is the first to state that heat and mechanical energy are two forms of the same thing.

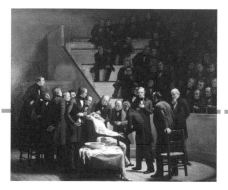

1. Geospiza magnirostris.
3. Geospiza parvula.
2. Geospiza fortis.
4. Certhidea olivacea.

1835 While a scientific officer on the *Beagle*, a British naval ship sent to survey the coast of South America, Charles Darwin visits the Galapagos Islands. Observes that the unique bird species on individual islands have many physical traits in common with each other. Later posits that closely related finches developed from a common ancestor on mainland of South America.

1837 French biologist Henri Dutrochet shows that only those parts of plants that contain the green pigment chlorophyll absorb carbon dioxide, and that they only do so in the presence of light. An important step in the understanding of photosynthesis: the transformation of light energy into chemical energy.

1837 Samuel Morse patents his version of the telegraph in America, a machine that sends letters in a code made of dots and dashes. Using the system of electrical relays invented by Joseph Henry, Morse is able to send a message over 10 miles of wire. Within 30 years, wires are strung across continents and under oceans, revolutionizing communication.

1839 Louis Jacques Daguerre, a French chemist and artist, announces a process for making photographs. Unlike modern photographs there are no negatives produced; rather the image is exposed directly onto a polished silver plate. The pictures are known as daguerreotypes.

1840 After conducting field studies in the Swiss Alps, Swiss geologist Louis Agassiz proposes his theory of Earth's "Ice Age," and describes the motions and deposits of glaciers. Points to boulders scattered across northern Europe and scratched rock surfaces as proof that enormous ice sheets once moved down from the Alps to cover most of Europe.

1839 "Cell theory" is first stated by German biologist Theodor Schwann and German botanist Matthias Schleiden. Proposes that both plants and animals are made up of cells, that eggs are cells, and that all life starts as a single cell. Lays the foundation of modern biology. Final step in cell theory is taken two decades later, when scientists realize that new cells come from the division of existing cells.

1842 Austrian physicist Christian Doppler proposes that the change in pitch of a sound is directly related to the velocity of the sound waves. The principle also applies to light and other electromagnetic waves. The Doppler effect can be heard when an ambulance passes with its siren on. As it approaches, sound waves are pressed together and the sound is higher in pitch; when it moves away, the waves are stretched out and the pitch drops.

1843 English astronomer John Adams studies the irregularities in the orbit of Uranus, and says there must be another, more distant planet that is exerting a gravitational pull on Uranus. Adams roughly calculates its position, and astronomers begin to search the skies for the new planet. It's sighted in 1846 by Johann Galle at the Berlin Observatory, and named Neptune.

1846 A measles epidemic occurs in the Faeroe Islands, and Danish physician Peter Panum studies how the disease progresses from person to person and island to island. Describes the disease's symptoms, incubation period, and infectious nature in one of the first works of epidemiology.

1848 William Thomson (later known as Lord Kelvin) proposes the concept of absolute zero, the lowest temperature possible: -273° Celsius, or -460° Fahrenheit. At that temperature, Thomson says, the energy of molecules would be zero, and the molecules would be absolutely still. While the concept has been accepted as true, absolute zero has never been reached.

1845 Christian Schönbein invents the synthetic chemical nitrocellulose, or guncotton, by accident. He wipes up a spilled nitric acid solution with his wife's cotton apron and hangs it up to dry, whereupon the apron explodes with a flash. Nitrocellulose is briefly used in firearms and leads to the invention of other explosives.

1847 James Prescott Joule states the first law of thermodynamics, also called the law of conservation of energy: that energy is neither created nor destroyed, but only changed from one form to another via work or heat. Demonstrates his theory with a paddle-wheel that stirs water vigorously, causing the water to heat up. Three years later, German physicist Rudolf Clausius states the second law of thermodynamics—that entropy always increases in a closed system.

Epidemiology and Public Health

In 1832 a cholera epidemic kills over 3000 people in London, and in the following decades disease outbreaks continue to sweep through Europe's densely populated cities. A few doctors go beyond treating their patients' symptoms, and wonder about how disease is spread. John Snow's work becomes a model for future researchers in the new field of epidemiology; they try to determine the cause and conduits of a disease by examining the risk factors patients were exposed to.

1851 French physicist Léon Foucault uses a pendulum suspended in a monumental Parisian building to demonstrate Earth's rotation on its axis. The pendulum

is suspended so that its swing direction does not change. Since its path on the floor changes during the course of a day, it demonstrates that the floor is moving due to the earth's rotation.

1852 English chemist Edward Frankland introduces the concept of the valence of elements to chemistry: that each atom can combine with a certain number of other atoms, and no more. It's later understood that atoms combine by sharing

electrons to form chemical bonds, and the number of spaces left in the atom's electron shells determine how many other atoms it can form bonds with.

1854 John Snow shows that sewage-contaminated drinking water causes cholera. After studying a previous outbreak, he determined

DEATH'S DISPENSARY.

that cholera couldn't be caused by noxious air because it affected the intestines, not the lungs. Then proves his theory by removing the pump handle of a London well contaminated with sewage, thereby dramatically reducing the number of cholera cases in the vicinity.

1855 German glassblower Heinrich Geissler makes the first dependable vacuum tube, which he then fills with rarified gas. When a current is passed through the tube, electrons are dissociated from the gas molecules, producing lighting effects. The device, used largely for entertainment, is soon modified by others to produce cathode rays: those tubes are eventually used in televisions and other electronics.

1856 Louis Pasteur reports his discovery that fermentation is caused by microorganisms (yeast). Previously, it was thought to be caused by  chemical reactions, and the microorganisms that were viewed in the liquid were thought to arise by spontaneous generation. Helps prove germ theory: that microorganisms are the cause of many diseases and processes.

1859 Charles Darwin's *On the Origin of Species* explains his principles of natural selection and evolution. Contradicting the Christian belief that each individual species was created by God, he states that all species evolve from preexisting species. Proposes "natural selection" as the evolutionary mechanism: individuals that are better adapted to the environment fare better and have more offspring, thus increasing the likelihood that their traits will be passed down. British biologist Alfred Russel Wallace independently arrives at similar conclusions, but Darwin's theory is more thoroughly supported by research.

Debate over Darwin

Darwin arrives at his theory of natural selection in the 1830s, but doesn't publish his ideas until 1859, in part because of their controversial nature. When his book is published, members of the Church of England denounce it for implying that new species evolve without assistance from a divine creator. Many members of the scientific establishment and much of the public are outraged, believing that Darwin's theories erase the distinction between humans and animals.

1857 Working in the garden of his Austrian monastery, Gregor Mendel starts the experiments with peas that will lead to his discovery of the laws of heredity. He cross-pollinates two strains of pea plants, one with purple and one with white flowers, and gets hybrids with all purple flowers. When he uses the hybrids to make a new generation, however, he gets both purple and white flowers. Explains that physical traits are inherited in dominant and recessive genes from both parents.

1859 German chemist Robert Bunsen works with German physicist Gustav Kirchhoff to invent the spectroscope, used to map and analyze the chemical makeup of a substance. When the substance is heated to incandescence, its component elements emit specific light frequencies that appear as sharply defined bands of color. These patterns are sometimes called spectral signatures. Spectroscopes are later used in astronomy to determine which elements make up distant stars.

Internal Combustion to the Atomic Bomb: 1859 to 1945

By the middle of the 19th century, millions of people in Western Europe and North America found their everyday lives changing rapidly, thanks to new technology and the rapid pace of industrialization. Workers, now more mobile, left the countryside and flocked to the cities for factory jobs. In London, for example, the population ballooned from 1 million in 1800 to 6.5 million by the century's end. People had barely adapted to the possibilities of steam-engine locomotives when the invention of the internal combustion engine in 1859 foreshadowed a new revolution in transportation. The Model T Ford was introduced in 1908, and a year later a French aviator made the longest flight yet, his airplane sputtering through the 37 minutes it would take to cross the English Channel.

The pace of scientific progress became exponential as the 19th century hurried on toward its end, due in part to better communication between scientists. It was a new world of radio broadcasts and transatlantic telegraphs; when the German physicist Wilhelm Roentgen announced his discovery of x-rays in December of 1895, the news spread rapidly around the globe. Two months later, an American doctor first used an x-ray to examine a patient's fractured bone. Soon x-ray machines appeared not only in doctors' offices, but also in circuses, where amazed patrons marveled at the sight of their own skeletons.

In the 20th century, it is said, more scientists lived than in all of previous history. The man who has become an emblem of scientific genius was just twenty-six years old and working as a patent clerk in 1905, when he sent three papers to the premier journal of physics, asking that they be published "if there is room." Albert Einstein's work fundamentally altered the way physicists viewed the nature of light, space, and time. In his special theory of relativity, he proposed that measurements of space and time are not absolute, but instead vary depending on the observer's location and motion. This explains why a passenger on a plane doesn't feel like he is traveling hundreds of miles per hour and can only get a sense of the plane's velocity by looking at the landscape below, or why we're not aware of the movement of the earth as it revolves on its axis and orbits the sun.

Scientists in this period gained new, sophisticated understandings of our world, from the vast expanses of outer space to the spaces within atoms. Before 1897, physicists believed atoms to be the smallest building blocks of matter. But in that year, the English physicist J. J. Thomson showed that atoms were made of still smaller components when he revealed the existence of the negatively charged particles now called electrons. That set off a decade of speculation over the structure and nature of the atom, with Thomson suggesting a "plum pudding model" of the atom, where a positively charged field was studded with electrons, like a pudding full of plums.

In 1909, the New Zealand-born physicist Ernest Rutherford

proved this theory wrong with his famous gold foil experiment. He fired a stream of particles at an extremely thin sheet of gold foil expecting that the particles would pass through it, as he believed the mass of the gold atoms was evenly distributed. Instead, about 1 in every 8,000 particles bounced back. The researchers were shocked. "It was almost as incredible as if you fired a fifteen-inch shell at a piece of tissue paper and it came back and hit you," Rutherford later said. He had proven that the atom was mostly empty space, with its mass concentrated in a dense, positively charged nucleus and electrons in orbit around the nucleus.

Physicists built on that understanding of atomic structure, combining it with the knowledge of radioactive decay gleaned from the work of Marie Curie and others who had shown that radioactive elements spontaneously break down, emitting energy in the form of smaller particles. In the man-made nuclear fission reaction first achieved in 1938, physicists "split the atom" by firing particles at a uranium atom. The particles that split off hit other uranium atoms, creating a chain reaction that released enormous amounts of energy in the process. By carefully controlling this chain reaction, physicists realized they could offer the world a cheap and plentiful source of energy, and therefore electric power; by maximizing the reaction, they could create bombs of unprecedented destruction.

After World War I, governments boosted their funding of scientific research, partly in hopes of military applications for work in chemistry, physics, and engineering. Nuclear weapons became a tantalizing goal for military leaders in the 1930s, as the world edged toward war. While physicists in a number of countries worked on nuclear fission, Germany lost its edge after Adolf Hitler was appointed chancellor in 1933. Hitler's Nazi Party began pushing forward a program of racial purity and anti-Semitism, and consequently lost many of Europe's most brilliant scientists. Jewish scientists in Germany lost their jobs, and many fled the country in fear, seeking safe haven in England or the United States.

Still, at the beginning of World War II, American physicists worried that Germany was racing to produce a nuclear bomb, pointing to the country's sophisticated laboratories and its control of uranium mines in Czechoslovakia. In 1939, Einstein gave voice to these concerns in a letter to President Franklin Delano Roosevelt, urging an initiative to build a bomb before Germany did so; Einstein would later say that signing the letter was the greatest mistake of his life. The resulting Manhattan Project was the most massive scientific undertaking ever attempted, bringing together the most brilliant physicists, chemists, and engineers in the country to labor for six years in strict secrecy. On July 16, 1945, they tested the first nuclear bomb in the desert of New Mexico; less than a month later American planes dropped nuclear bombs on the Japanese cities of Hiroshima and Nagasaki to end World War II.

The bombs killed over 110,000 people in the two cities instantly, with thousands more dying of radiation poisoning in the following months. Many of the brilliant scientists who took part in the bombs' design and production were appalled by what they had wrought, including Einstein: "It has become appallingly obvious that our technology has exceeded our humanity," he said.

1859 Belgian inventor Jean Joseph Étienne Lenoir develops the first functional internal combustion engine, using coal gas as fuel. His engines are primarily used as stationary power sources, to drive printing presses and water pumps. A decade later, an Austrian inventor puts an internal combustion engine on a cart, leading to the first commercial automobiles being marketed as "horseless carriages" in the 1890s.

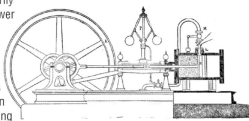

185,000 miles/second

1862 The most accurate measurement of the speed of light to date is published by French physicist Léon Foucault, who uses a rotating mirror instrument, an improvement on the rotating cog device used in the 1840s by Hippolyte Fizeau. Foucault's measurement of 185,000 miles per second is only 0.6% off from the current accepted value of 186,282 miles per second.

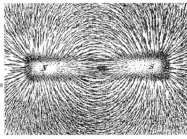

1860

1861 French physiologist Paul Pierre Broca conducts an autopsy of a man who couldn't speak intelligibly, a patient who had been nicknamed "Tan," because that was the only word he could say. Broca sees a lesion in the man's brain, on the left frontal lobe. Demonstrates the concept of cerebral localization: that particular regions of the brain are connected to particular faculties, in this case, speech. This part of the brain is now called Broca's area in his honor.

1862 At the 1862 London International Exhibition (pictured), English inventor Alexander Parkes displays many items made from a derivative of cellulose nitrate. The new material, which he calls Parkesine, is the first man-made plastic. While the company Parkes begins isn't a success, his work anticipates and encourages further use of plastics.

1864 Scottish physicist James Clerk Maxwell publishes a book introducing the concept of an electromagnetic field. He shows that when a magnet moves it produces an electric current, and when an electric current flows, it causes magnetism. The interaction of the two forces produces waves, or electromagnetic radiation. Maxwell explains that visible light is one form of this radiation.

1866 German physicist August Kundt devises a method for measuring the speed of sound in different gases. He sends sound waves through a gas-filled tube where a fine dust is scattered inside, and analyzes the wave patterns formed in the dust to determine how fast the waves travel. Realizes that the speed of sound varies according to which gas the sound waves are passing through, and according to temperature and pressure. Traveling through air, the speed of sound is around 1100 feet per second.

1867 Building on Louis Pasteur's germ theory, which states that microorganisms cause rotting and fermentation, British surgeon Joseph Lister applies lessons to hospitals and operating rooms. Publishes a series of articles, "Antiseptic Principle of the Practice of Surgery," which advocate the use of carbolic acid to sterilize surgical instruments and to kill bacteria in wounds. (His carbolic spray shown here on a British postage stamp.)

1867 Alfred Nobel, a Swedish chemist and armaments manufacturer, patents a relatively safe and stable form of the explosive nitroglycerine, which he calls dynamite. He makes a fortune on this invention and others, and founds the Nobel Prizes in 1895 in an attempt to improve his legacy to the world. Three scientific prizes are given for breakthroughs in physical science, chemistry, and medical science, while two others are awarded for literature and promoting peace. Later, a prize in economics is added.

Medical Conditions

While many advances had been made in understanding disease and public health by the mid-19th century, physicians are still hampered by some very wrong ideas. Many believe in the miasma theory: that foul-smelling vapors, like those found in swamps and sewers, are filled with particles that cause illness. Lister's suggestion that surgeons wash their hands with a carbolic acid solution before operating is considered foolish by many who believe that diseases are spread through the air, not through human contact. Gradually, however, Pasteur's germ theory is proven through experimentation.

1869 Russian chemist Dmitri Mendeléev publishes his first version of the periodic table of elements. Classifies the elements according to their chemical properties, and arranges them in order of atomic mass. Mendeléev leaves gaps in his table, predicting the discovery of several elements with specific atomic masses.

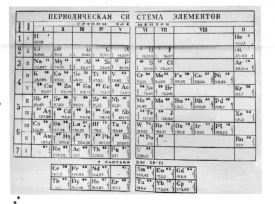

1876 While working in Boston, Scottish-born inventor Alexander Graham Bell patents the telephone, founds the Bell Telephone Company one year later. Sound waves produced by human speech vary the electrical current that passes through the wire, and a receiver on the other end reproduces those variations in audible form. First words spoken on the telephone are from Bell to his assistant: "Mr. Watson—come here—I want to see you."

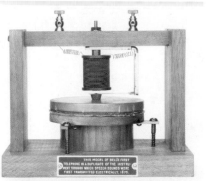

1871 In his second book on evolutionary theory, Darwin details his theory of sexual selection. Explains that features which don't give an animal an obvious advantage in competing for resources—like a male peacock's ungainly tail—serve the purpose instead of attracting a mate. Individuals that have more success in attracting mates are likely to pass their genes on to more offspring.

1874 German psychologist Wilhelm Wundt publishes his seminal work, *Principles of Physiological Psychology*. Argues that to understand human perceptions, emotions, motives, and ideas, social scientists should carry out careful observations under controlled conditions. The first advocate of experimental psychology, which uses experimental methods drawn from the natural sciences.

1879 Louis Pasteur discovers by accident that weakened cholera bacteria don't cause fatal disease in chickens, and that those chickens are thereafter immune to the disease. While Edward Jenner had previously made a similar discovery with smallpox, this makes the general principles of vaccination clear. Pasteur goes on to develop vaccines against anthrax and rabies.

1879 Thomas Edison in America and Joseph Swan in England independently develop electric light bulbs that burn for practical lengths of time. Previous inventors had produced electric lighting in laboratory conditions, but their devices weren't suited for widespread commercial distribution. The electrical current in Edison's light bulb runs through a thin filament of carbon in a vacuum.

Louis Pasteur

Few scientists do as much for public health as the French chemist Louis Pasteur. His experiments prove the germ theory of disease, which explains that microorganisms are responsible for infection and decay, and changes medicine forever. He invents the process that is later named pasteurization in his honor, where liquids such as milk are heated to kill the bacteria present. Finally, his interest in diseases leads him to develop vaccines against three powerful killers.

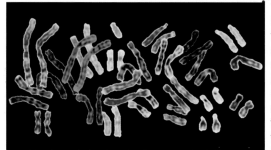

1882 Walther Flemming, a German biologist, makes the first detailed observations of chromosomes, thread-like structures in a cell nucleus. He observes the process of cell division, during which chromosomes are split into two identical halves, and names the cell division process mitosis. Shows that all nuclei arise from the division of preexisting nuclei.

Germany Leads the Way

After prime minister Otto von Bismarck unifies Germany and founds the German Empire in 1871, the nation enters into a period of peace, stability, and prosperity. Economic and social reforms push Germany into position as the world's leading industrial power, with its textile and metal industries supplying the western world. Nearly universal literacy and a vibrant intellectual atmosphere produce scholars who make exciting discoveries in numerous scientific fields.

1882 First commercial hydroelectric power plant goes into operation on the Fox River in Appleton, Wisconsin. The river is dammed, and water is then pushed through to drive a turbine and generate electricity. At first it produces only enough electricity to power two paper mills and lights for a private home.

1880

1886 Nitrogen-fixing bacteria discovered by two German agricultural chemists, Hermann Hellriegel and Hermann Willfarth. These beneficial bacteria (such as Rhizobium bacteria, pictured) live on the roots of bean plants and convert atmospheric nitrogen into a form plants can use for growth. Crops grown in a field where beans have recently been grown benefit from the bacteria on the decaying bean roots, leading to increased crop yields.

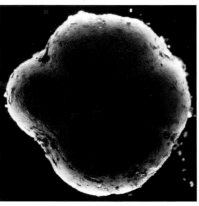

1882 The Russian-born microbiologist Ilya Mechnikov proposes that certain white blood cells are the basis for the immune system's response to disease. He calls these white blood cells phagocytes (from the Greek for "devouring cells"), and explains that they engulf and destroy harmful bacteria in a process he calls phagocytosis.

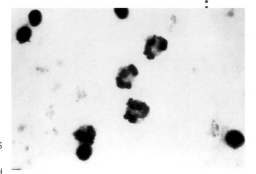

1888 German physicist Heinrich Hertz produces and detects radio waves for the first time. His experiments with these electromagnetic waves show that waves can be sent from sender to receiver. Hertz thinks his discovery has no practical application, but within a decade the wireless telegraph (telegraph receiver pictured) and the radio are invented.

1889 The Italian Camillo Golgi and the Spaniard Santiago Ramon y Cajal independently study the nerve cells of the brain, which Cajal calls neurons. Cajal theorizes that the nervous system is composed of billions of separate neurons that communicate by sending chemical impulses across tiny gaps called synapses. (One of his drawings is pictured.) The work of the two men marks the beginning of modern neuroscience.

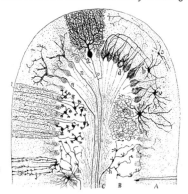

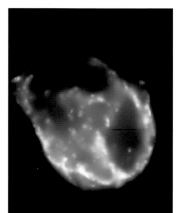

1892 A bright nova appears in the constellation Auriga, and American astronomer E. E. Barnard observes that it is emitting a cloud of gas. First clear evidence that novas are exploding stars. Eventually, astronomers understand that a nova is a nuclear fusion reaction that blows gases away from the star's surface in a burst of light.

1891 Maximilian Wolf, a German astronomer, is the first to use time-exposure photographs to discover asteroids, the small planetoids in our solar system that orbit the sun. A vast improvement on the discovery method used since 1801 of charting the stars on repeating nights and looking for stars that had "moved." In Wolf's photographs, asteroids show up as streaks of light against the stars' fixed points of light.

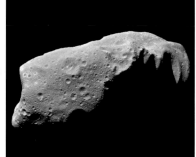

1892 Russian scientist Dmitri Ivanovski shows that a disease that affects tobacco plants is caused by an organism too small to be seen by microscope, which can pass through a very fine filter that screens out bacteria. He calls the organism a virus, and the field of virology is born. A decade later, yellow fever will be recognized as the first known human viral disease.

New Waves

When James Clerk Maxwell introduced the concept of electromagnetic radiation in 1864, he gave physicists a new challenge: to discover new waves with wavelengths longer and shorter than the visible light spectrum. For scientists, electromagnetism was an exciting new way to understand previously puzzling phenomena such as light and heat, and they eagerly took up the search to see what else could be discovered. The first to be found are radio waves, which have a much longer wavelength than visible light. Within 10 years, the short-wavelength x-rays are found.

1893 Friedrich Wilhelm Ostwald gives the first modern definition of a catalyst: a substance that speeds up a chemical reaction but isn't itself transformed or consumed. In 1902 he patents a catalytic reaction he calls the "Ostwald process;" this system for producing nitric acid becomes very useful in the manufacture of explosives and fertilizers. Ostwald, a Russian-born chemist who works in Germany, is considered the founder of modern physical chemistry.

1897 English physicist J. J. Thomson discovers the electron, the first known particle smaller than an atom. Inspired by Roentgen's discovery of the x-ray, Thompson studies electrical charges in a cathode ray tube, and determines that negatively charged particles (electrons) are the carriers of electricity.

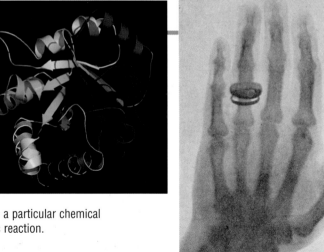

1894 German chemist Emil Fischer studies the enzymes (or natural catalysts) that convert sugar to alcohol. Notes that each kind of enzyme is structured to recognize only one particular form of the sugar compound, using what he calls a lock and key mechanism. Indicates that each living cell contains a multitude of enzymes, each designed to recognize a particular chemical compound and bring about a specific reaction.

1895 During an experiment on cathode rays, German physicist Wilhelm Konrad Roentgen discovers that invisible rays emanating from the tube cause a fluorescent effect on a piece of cardboard painted with a chemical. Calls this new type of electromagnetic radiation x-rays after the mathematical term for an unknown. When he puts his hand between the tube and the cardboard, he sees the bones of his hand in the first radiographic image. Less than three months later, first x-rays are used by American doctors to diagnose patients.

1898 French chemist Marie Curie and her husband Pierre discover that uranium gives off "uranium rays," which Marie renames radioactivity. When the Curies refine pure radium—one of the elements they discover—it emits an eerie blue glow showing that radioactive elements spontaneously emit radioactive energy. Marie becomes the first woman appointed to teach at the prestigious University of Paris, and is the first woman to win a Nobel Prize.

Radiation's Dark Side

Marie Curie is unaware the element uranium that enchants her with its blue glow is slowly poisoning her. It's not until many years later that scientists learn that high doses of radiation interfere with the division of cells in the gastrointestinal tract, and cause genetic mutations that can lead to cancer. Curie dies in 1934 from leukemia, almost certainly acquired from her prolonged exposure to radiation during her scientific research. Until recently, her lab notes were too radioactive to be safely handled.

1900

1898 While studying malaria-stricken birds in India, English physician Ronald Ross locates the malaria parasite in a mosquito's salivary glands, and proves that the mosquito transmits the parasite from one bird to another. Shows that this must also be the disease's mode of transmission in humans. Previously, doctors thought the disease was caused by something in the air in the swampy and damp environments where malaria was common.

1900 German physicist Max Planck explains in a paper that electromagnetic energy doesn't flow continuously, but rather in chunks, or "quanta." This rebuts a theory of classical physics and marks the beginning of quantum physics, one of the defining scientific fields of the 20th century.

1901 Working together at McGill University in Canada, Ernest Rutherford and Frederick Soddy discover that the radioactive element thorium, left to itself, changes into another element. They call this process radioactive decay, and show that it happens when unstable atomic nuclei spontaneously give off subatomic particles. Soddy later shows that the decay of a uranium atom (pictured) produces radium, and that radium decays to helium.

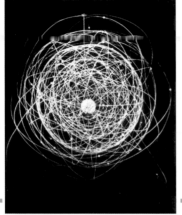

1903 On December 17, American brothers Wilbur and Orville Wright launch their first airplane, the Wright Flyer, at Kitty Hawk, North Carolina. Each brother flies the airplane twice, with Wilbur's last flight being the longest and only truly controlled flight of the day, traveling 852 feet and lasting 59 seconds.

The Flying Brothers

Mechanically talented from a young age, the Wright brothers leave high school to open a bicycle repair and manufacturing company. They use the proceeds from the business to fund their research into airplanes, eventually building a wind tunnel to test over 200 different wing shapes. Their fascination with flight dominates their lives; neither brother ever marries, and they move together to Kitty Hawk, North Carolina to continue their experiments in seclusion and to benefit from the region's strong winds. They apply for a patent for their "flying machine" in 1903, but can't get a contract to build planes for the U.S. military because officials don't believe their claims.

1901 Austrian physician Karl Landsteiner explains why blood transfusions often cause fatal reactions in patients with his definition of the four blood groups: A, B, AB, and O. The different groups are defined by the presence or absence of certain proteins on the surface of the red blood cells. When transfusion patients receive a different type of blood from a donor, their bodies' treat the incoming red blood cells as foreign bodies and destroy them, with dangerous side effects.

1902 British physiologists Ernest Starling and William Bayliss identify secretin, a hormone secreted by the small intestine that stimulates pancreatic secretions to aid digestion. The first recognition of hormones, the chemical messengers that help regulate the body by moving through the blood stream to a target organ, tissue, or clump of cells.

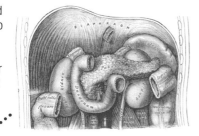

1905 German-born physicist Albert Einstein publishes four papers in the same year, while working as a Swiss patent clerk. Three of those papers are considered revolutionary. The first describes his theory on Brownian motion, the random movements of minute particles. The second is on the particle (quantal) nature of light. The last explains his special theory of relativity, which states that the speed of light is constant and contains the famous formula, $e=mc^2$, to describes the relationship between mass and energy.

$$e=mc^2$$

Einstein's Miracle Year

Albert Einstein is in his mid-twenties and working as a clerk in the Swiss patent office in 1905, when he publishes three papers that change the course of physics and science. Commonly called the *Annus Mirabilis* Papers (from the Latin meaning "year of wonders"), the most startling of the three states his special theory of relativity and upsets the standard view of space and time. He rejects Isaac Newton's laws of mechanics, which assume the existence of absolute space and absolute time, and instead shows that measurementsof space and time vary depending on the observer's location and motion.

1904 The Russian physiologist and physician Ivan Pavlov shows that dogs begin to salivate not only when they see food, but also when exposed to a stimulus they've come to associate with food, like a ringing bell or whistle. This gives birth to the theory of conditioning, which suggests that humans and animals learn through experience to have uncontrolled emotional and physical responses to certain stimuli.

1906 Prolific Canadian inventor Reginald Fessenden invents AM radio and makes the first audio transmission via radio waves from Brant Rock, Massachusetts on Christmas Eve. Heard by ships off the coast, Fessenden's transmission includes the reading of a passage from the Bible and music, his own violin rendition of "O Holy Night." Previously, the only radio transmissions made were of Morse code signals.

1907 Thomas Hunt Morgan begins his work with fruit flies that will prove that inherited physical traits are determined by genes, the basic units of heredity that are carried on chromosomes. After reading the recently rediscovered theories of inheritance proposed by Gregor Mendel, Morgan crossbreeds white-eyed and red-eyed flies. He finds that all the offspring have red eyes, suggesting that the red-eye trait is dominant. That means when an offspring inherits both genes, one from each parent, the red-eye trait will be expressed.

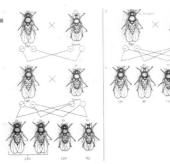

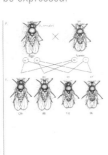

The Humble Fruit Fly's Role in Science

Because of Morgan's experiments in his famous "Fly Room" at Columbia University, fruit flies become common test subjects in genetic experiments. They're useful because they produce a new generation every two weeks, allowing for easy study of inherited traits, and because only 4 pairs of chromosomes carry all their genes. It's later determined that about 60 percent of genes that cause human genetic disorders have a recognizable match in fruit flies' genetic code, making them a useful model organism for the study of those diseases.

1907 American chemist Bertram Boltwood theorizes that lead is the stable end product of uranium decay. Realizing he can determine the age of a rock if he knows uranium's rate of decay, he studies the proportions of uranium and lead in a mineral formation. Dates the formation of a Connecticut mineral to 410 million years ago. A step toward establishing the accurate age of the earth at a time when many people believed Judeo-Christian estimates of a 6000-year-old planet.

1910 German doctor Paul Ehrlich creates a synthetic arsenic compound that kills the bacteria that causes syphilis without harming the human host. Sets off a search for other synthetic drugs that are called "magic bullets" because they specifically target and kill the disease-causing organism.

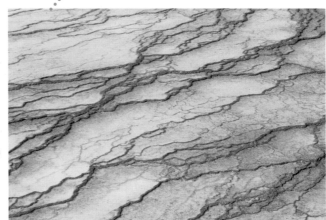

1910 English physicist J. J. Thomson measures the atomic masses of neon and finds 2 forms that have different masses, neon-20 and neon-22. The first recognition of isotopes: different forms of an element that have different atomic masses. Isotopes had been previously observed by other scientists, but they were mistakenly believed to be new elements.

1911 Dutch physicist Heike Kamerlingh Onnes discovers that when certain metals, such as mercury, are cooled to near absolute zero, their resistance to the flow of electricity drops to nothing. He calls this state superconductivity, and while he doesn't understand how it works, he guesses that it could be very useful, as the electric current carried by superconductors loses no energy through resistance and dissipation.

1912 After studying similar plant fossils from South America, Africa and Australia, German meteorologist and explorer Alfred Wegener first proposes his theory of continental drift. He believes all continents joined together about 300 million years ago to form the ancient supercontinent of Pangaea: from Greek words for "all earth." Proposes that Pangaea eventually divided into today's continents which migrated around the globe as the tectonic plates shifted.

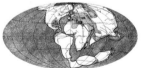

1911 New Zealand-born physicist Ernest Rutherford presents his theory of atomic structure, consisting of a dense, positively charged nucleus at the center of the atom, surrounded by negatively charged electrons. He comes to the realization after working on radioactive elements for years, and forming the "disintegration theory" of radioactivity, which views radioactivity as the breaking apart of atoms into their component parts. His device for studying the disintegration of elements is pictured.

1912
Henrietta Leavitt, a deaf astronomer working at Harvard, finds a way to calculate the inherent brightness of stars, a measure of how much energy they are emitting. By then measuring their apparent brightness visible to observers once the light has traveled through space to our planet, Leavitt calculates the distance of stars. Other researchers use this procedure to prove that galaxies such as the Andromeda Galaxy are outside the Milky Way, and to measure our galaxy's size.

1912 English biochemist Frederick Hopkins shows that "accessory food factors," later named vitamins, are essential for growth and health. These proteins aren't manufactured by the body, and need to be acquired through diet. Other researchers recognize that diseases such as scurvy, rickets, and beriberi are caused by vitamin deficiencies.

NAME	CHEMICAL NAME	DEFICIENCY DISEASE
Vitamin A	Retinol	Night-blindness, Keratomalacia
Vitamin B^1	Thiamine	Beriberi
Vitamin B^2	Riboflavin	Ariboflavinosis
Vitamin B^3	Niacin	Pellagra
Vitamin B^9	Folic acid	Birth defects (if deficient in pregnant women)
Vitamin B^{12}	Cyanocobalamin	Pernicious anemia
Vitamin C	Ascorbic acid	Scurvy
Vitamin D^1-D^4	Lamisterol, Ergocalciferol, Calciferol, Dihydrotachysterol, 7-dehydrositosterol	Rickets

The Atom

Following J. J. Thomson's 1897 discovery of the first subatomic particle, the electron, the race is on to determine the components and structure of the atom—which was previously believed to be the smallest building block of matter. Rutherford advances the field when he discovers the nucleus, and uses his famous gold foil experiments to demonstrate that most of the atom's mass is concentrated in the nucleus. He fires particles at a thin gold foil and determines how many come out the other side. Finds that only 1 in 8000 particles are deflected, the rest passed through the atom's empty spaces surrounding the nuclei.

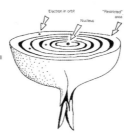

1913 Danish nuclear physicist Niels Bohr expands on Rutherford's atomic model with his theory of electron orbits. States that rather than being sedentary, electrons move around the atomic nucleus in fixed orbits, and the number of electrons orbiting a nucleus determines an element's chemical properties. He also states that the electrons give off or absorb fixed amounts of energy (quanta) by jumping from one orbit to another. An important step in the new field of quantum mechanics.

1913 At his Highland Park, Michigan, factory, American car manufacturer Henry Ford develops the first factory assembly line using a moving conveyer belt. Based on the assembly lines used by the massive meat markets in Chicago, Ford's assembly line vastly increases the efficiency of production: construction of a Model T car takes 1.5 hours as compared to over 12 hours under the old system.

The Earth's Structure

In 1914 German geologist and seismologist Beno Gutenberg helps describe the earth's structure when he discovers a discontinuity in the behavior and velocity of earthquake waves to a depth of about 1860 miles beneath the earth's surface. Declares that this marks the boundary between the earth's mantle and its core, which is made of iron and other dense elements.

1915 Albert Einstein publishes his theory of general relativity, stating that the flat expanse of the space-time continuum is curved by the gravity of objects, and that even light is bent by gravity. Leads to the explanation of black holes, dense concentrations of mass with such a strong gravitational force that nothing— not even light—can escape them.

20th Century Technology

The pace of technological advancement stuns the citizens of America and Western Europe, where the most rapid changes take place. In the first decade of the 20th century, radio programming becomes popular, and the first silent films are distributed. The Model T Ford moves quickly from a coveted status symbol to a rather common convenience: during the 19 years of Model T production, over 15 million are produced. Information, goods, and people all travel faster through the world. In Carl Sandburg's 1916 book *Chicago Poems*, he captures the city's rush of industry: "She could see the smoke of the engines get lost down/ where the streaks of steel flashed in the sun and/ when the newspapers came in on the morning mail."

1915

1915 The first American transcontinental telephone call is made between Bell in New York City (pictured more than a decade earlier calling only as far as Chicago) and Bell's former assistant Watson in San Francisco. The phone lines were strung across the country by the American Telephone and Telegraph Company, later known as AT&T.

1916 American chemist Gilbert Lewis explains the chemical bonds that hold molecules together with his theory that electrons are shared by adjacent atoms. He calls his idea of shared electrons "covalent bonding" and coins the term "odd molecule" to describe a molecule with an odd number of electrons.

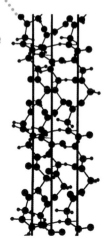

1917 Dutch astronomer and physicist Willem de Sitter demonstrates that Einstein's theory of general relativity proves that the universe is expanding. A theoretical model that carries his theory to an extreme—where the universe expands exponentially until matter and light are diluted and unobservable—is later termed a "De Sitter Universe" after this discovery.

1921 Vilhelm Bjerknes, a Norwegian physicist and meteorologist, proposes that hot and cold masses of air interact to form cyclones and other weather events, and that most weather activity occurs on the edges of what he calls warm and cold fronts. Describes the movement of air masses as a function of pressure, temperature, and composition, but his complicated equations aren't frequently used until computers make them easier to solve.

1920

1920 British astrophysicist Arthur Eddington explains stellar structure, deducing that stars derive their energy by an ongoing process of nuclear fusion which transmutes hydrogen to helium in the heat and pressure of their cores. Previously, astrophysicists theorized that stars gained energy by consuming dust and comets that fall into their gravitational pulls. Later, the hydrogen bomb will duplicate the fusion reaction that takes place in stars.

1921 Canadian scientist Frederick Banting, with the aid of his student Charles Best and in the lab of Professor John Macleod, extracts the hormone insulin from a pancreas and starts experimenting on ten dogs in an effort to treat diabetes. Although researchers have guessed since the 19th century that the pancreas secretes a substance that regulates blood sugar levels, this is the first successful attempt to treat diabetes with insulin (pictured).

1924 German psychologist Hans Berger develops the electroencephalogram (EEG) to measure electrical activity in the brain, or "brainwaves" (the first EEG, pictured). Electrodes are placed on the scalp, which measure the voltage traveling between them. Later used to diagnose and study brain damage, epilepsy, and sleep disorders.

1921 Working independently, English neuroscientist Henry Dale and German pharmacologist Otto Loewi discover that certain chemicals stimulate nerves to fire —rebutting the notion that neurons communicate with each other purely through electrical signals. They develop the theory of the chemical transmission of nerve impulses, and Dale identifies the first neurotransmitter, acetylcholine.

1924 From an observatory in southern California, American astronomer Edwin Hubble proves that other galaxies exist as true independent systems outside our Milky Way system by determining the distance of the Andromeda nebula (pictured) to be a million light years away. Explains to the public the vastness of the universe, and that our Milky Way is only one of a multitude of galaxies.

1925 A German oceanographic expedition, designed to recover gold from the oceans to pay Germany's war debts, is equipped with sonar to measure ocean depths. This technology, developed during World War I to help detect submarines, sends out sound waves, and measures the time it takes for the waves to bounce back from an underwater object. The German crew makes the first observation of the Mid-Atlantic Ridge, the underwater mountain range that runs from Iceland to the Antarctic Circle.

1926 American geneticist Hermann Muller, known for his work in spontaneous gene mutation, discovers that x-rays can induce genetic mutations. He increases the mutation rate of the fruit fly by a factor of 150 by irradiating them with x-rays. It's later discovered that other forms of radiation also cause mutations. (Pictured, mutated cockroaches.)

1927 German physicist Werner Heisenberg states the uncertainty principle of quantum mechanics, which describes the actions of subatomic particles. Explains that it is impossible to accurately determine two variables, such as position and momentum, of a particle simultaneously. Says that it's more useful to think of electrons as existing in a probability field, and that they only appear in a specific location when an observer tries to measure their location.

The Dawn of Genetic Engineering

By the 1920s, scientists understand that the basic units of heredity are genes, which are carried on chromosomes. When Hermann Muller discovers that he can change genes by inducing mutations, he shows researchers a new way to manipulate and study organisms, with major benefits for science. Eventually, advanced techniques allow researchers to "knock out" or turn off a single gene, allowing them to determine that gene's function.

1925 Australian-born anthropologist Raymond Dart announces his discovery of a hominid fossil that is neither human nor ape, and names the species *Australopithecus africanus*. Makes the identification from the skull of the so-called Taung Child, found in a limestone quarry outside of Taung, South Africa. Shows the species had a smaller brain than modern humans, but similar dental patterns and upright posture. Dart makes the contentious claim that the species represents the "missing link" in the evolution from apes to humans.

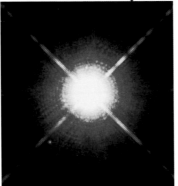

1927 Belgian astronomer and Roman Catholic priest George LeMaître proposes that the universe was created by the explosion of a dense concentration of matter and energy which he called the "cosmic egg" or "primeval atom," the first version of the currently accepted Big Bang theory of the universe's origins. Also suggests that galaxies are still moving outward propelled by the force of that initial explosion.

1927   Flying a single-engine airplane called *The Spirit of St. Louis*, American aviator Charles Lindbergh makes the first nonstop solo flight across the Atlantic. Lindbergh takes off from an airfield on Long Island, New York, and lands in Paris, France 33.5 hours later, where he is met by cheering crowds.

1928   Scottish pharmacologist Alexander Fleming discovers penicillin when he notices that a blue-green mold growing on a Petri dish inhibits bacterial growth. Due to the difficult nature of isolating penicillin, it isn't manufactured in mass quantities until World War II. The first modern antibiotic, it has an enormous effect on medicine and public health.

Wave-Particle Duality

Scientists thought they understood the nature of light at the end of the 19th century—it was a wave, a form of electromagnetic radiation. Then Albert Einstein published a paper showing that it can also be understood as a collection of particles, which he called photons. In the strange field of quantum mechanics, quantum entities are both waves and particles, and exhibit qualities of either form depending on which is being looked for and measured. This realization furthers physicists' attempt to understand the nature of matter at the smallest level, although it doesn't make intuitive sense.

1930   Cornelius Van Niel, a Dutch-American biologist, studies green algae and works out the chemical process of photosynthesis. Shows that carbon dioxide and water, in the presence of light, are converted to chemical energy for the plant and a waste product, oxygen.

1931 Indian-American astrophysicist Subrahmanyan Chandrasekhar, known widely as Chandra, studies white dwarfs: the small, dense, and very hot astronomical objects that result when a small star uses up the hydrogen that drives its nuclear reactions, and dies. Determines that beyond a certain solar mass, the density of a white dwarf causes it to collapse into a neutron star or black hole. Otherwise, they gradually radiate out their remaining energy and cool down.

1931 At the University of California in Berkeley, physicist Ernest Lawrence develops the first workable particle accelerator—the cyclotron. Generates the very high-energy particles needed to bombard an atom and cause it to fragment, by accelerating the particles around a circle. His original working prototype is made from electrical wire and sealing wax, and costs less than $30 to produce.

1931 George Washington Bridge opens over the Hudson River, linking New York City and New Jersey. It is the longest suspension bridge built to date, with a main span of 3500 feet long, nearly twice the length of the previous record-holder.

1932 English physicist James Chadwick discovers a new subatomic particle, the neutron, which has no electric charge, and forms part of the dense nucleus of the atom. Explains the nature of isotopes, the atomic versions of an element that have different atomic masses: their nuclei have the same number of protons, but different numbers of neutrons.

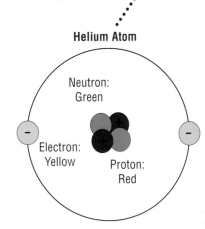

Helium Atom

Neutron: Green

Electron: Yellow

Proton: Red

1934 At the Dupont Chemical Company, American chemist Wallace Carothers experiments with polymers, long chains of repeating molecular units. He carries out the experiments that lead to the creation of a synthetic fiber soon named nylon, but dies before the mass production of nylon ushers in the synthetic fiber revolution.

1935 Portuguese neurologist Antonio Moniz develops the now discredited prefrontal lobotomy as a treatment for mental illness, in which the prefrontal cortex at the front of the brain is destroyed or disconnected. The technique is at first hailed as a powerful cure, as dangerous and disturbed patients become passive. By the 1950s, doctors begin to criticize the irreversible technique, as it works by destroying much of the patient's higher consciousness.

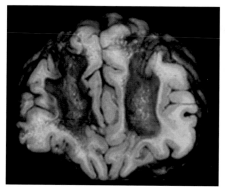

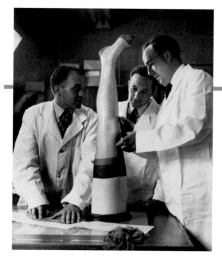

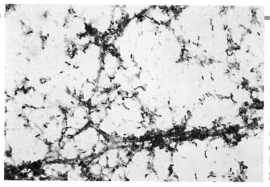

1935 German bacteriologist Gerhard Domagk uses his first sulfa drug, Prontosil, on his youngest daughter to prevent the amputation of a limb from a strepto-coccal infection (strep bacteria, pictured). Sulfa drugs, which are antibiotics derived from sulfonic acid, soon become famous as the first "wonder drugs." During World War II, American soldiers carried packets of sulfa powder with them to sprinkle on open wounds.

The Power of the Neutron

When James Chadwick announces the existence of a the neutron, he had no idea that knowledge of the subatomic particle's nature would allow both unprecedented destruction in the form of the atomic bomb, and cheap electric power in the form of atomic energy that would improve many lives. The neutron is the key to the nuclear chain reactions that take place in both nuclear weapons and nuclear power stations. When one atom is split, neutrons fly out and hit other atoms, causing them to break up and send out neutrons in a self-sustaining reaction, with the number of atoms involved increasing exponentially.

1936 The last known Tasmanian wolf dies in a zoo in Tasmania. The species was considered a pest and a threat to livestock, and was killed off by farmers and trappers who wanted to collect the government bounty. While many species had gone extinct in mainland Australia during European settlement, the extinction of the Tasmanian wolf is a milestone that encourages governments to protect endangered species.

1937 At the University of Palermo, Italian physicist Emilio Segrè discovers the first artificial element, in residue from a cyclotron experiment. Names it technetium after the Greek word *technètos*, meaning "artificial." Like many other Jewish scientists of his time, Segrè is soon forced to leave his home country for America because of anti-Semitic laws passed by Italy's hostile fascist government.

1936 In his book *The Origins of Life on Earth*, Russian biochemist Alexander Oparin proposes that life first evolved in the ocean, where a "primeval soup" of biochemical elements provided the building blocks for early life forms. He posited that through random chemical processes, atoms joined together into molecules, which joined into amino acids (pictured).

1937 German born biochemist Hans Krebs discovers the Krebs cycle, the cycle of oxidation and energy production in living cells. A series of chemical reactions, the cycle helps break down proteins, carbohydrates, and fats into their component elements, and stores the energy produced by the reactions.

Predicted Elements

When Dmitri Mendeléev developed his periodic table of elements in 1869, he left gaps in the orderly procession of atomic numbers where he expected newly-discovered elements to fit. He left one such space between atomic numbers 42 and 44, predicting the discovery of an element with chemical properties similar to the element manganese. While scientists eagerly searched for the predicted element for 60 years, they were unsuccessful because they lacked the technology: technetium, the 43rd element, is only produced in nuclear reactions.

1938 American behavioral scientist B. F. Skinner shows that rats and pigeons can be trained to perform complex tasks with positive or negative behavioral reinforcement—food or electric shocks. These experiments lay the groundwork for the controversial wing of psychology, behavior analysis.

1938 German physicist Otto Hahn fires neutrons at a piece of uranium, and is the first to split its atom. Soon realizes that this nuclear fission reaction produces enormous energy, and that a self-sustaining chain reaction can be created wherein neutrons that are split off from an atom smash into another atom, causing it to split. His experiments are based in part on the work of Lise Meitner, a female Jewish scientist who fled Germany shortly before Hahn's final round of experiments.

Lise Meitner, Unsung Heroine

While her colleagues know Lise Meitner as a brilliant physicist, like many other female scientists, she doesn't receive the wider acclaim her work warrants. She collaborates for over 30 years with Otto Hahn, and evidence shows that Hahn believes nuclear fission to be impossible until Meitner convinces him otherwise. However, Hahn alone receives the Nobel Prize in 1944 for the discovery.

1938 British engineer G. S. Callendar publishes an article stating that the burning of fossil fuels by humans is causing an increase in the amount of carbon dioxide in the Earth's atmosphere, and therefore affecting the climate. Using meteorological records from around the world, he determines that temperatures have trended upward during the first 40 years of the 20th century. An early—and largely ignored—statement of the greenhouse effect.

1939 The insecticide DDT is discovered by Swiss chemist Paul Muller, and is shown to be terrifically effective against insects like lice, mosquitoes, and flies that carry such diseases as typhus, malaria, and dysentery. It's hailed as a public health miracle and produced in mass quantities until 20 years later, when environmentalists show that it has toxic effects on other animals, including birds of prey.

1939 German-American physicist Walter Elsasser suggests that the Earth's core, made of liquid iron, has currents that maintain the planet's magnetic field. As the Earth turns and the liquid iron moves through already existing magnetic fields, it creates electric currents that spread out and create new magnetic fields, in a self-sustaining "dynamo" process. The magnetic field makes compasses function, and shields the earth from certain types of radiation from the sun by deflecting particles.

1939 At Iowa State University, engineers John Atanasoff and Clifford Berry demonstrate their prototype computer that is considered the model for all later electronic computer designs. The device, which weighs about 700 pounds, uses binary numbers to represent all data, and can solve linear equations. During WWII, the U.S. army develops an improved version (pictured).

1940 American biologist Herbert Evans proves that the chemical element iodine is used by the thyroid gland and is essential for human health. Iodine is needed for the thyroid's regulation of metabolism and growth; when it's not present in the diet, children's growth and brain development are severely impaired. Within 20 years, salt is laced with iodine to prevent deficiencies.

1941 American geneticists George Beadle and Edward Tatum demonstrate that genes control chemical reactions in cells. In their experiments, they trigger genetic mutations by exposing bread mold to x-rays, and then observe that the mutations cause changes in specific enzymes involved in cellular metabolism. Their discovery is popularly known as the "one-gene, one-enzyme" hypothesis.

1940

1940 Russian-American engineer Igor Sikorsky constructs the first successful helicopter designed for mass production, the Vought-Sikorsky 300. Though not the first helicopter, its simple and stable design, with a single main rotor and supplemental tail rotor, makes it suitable for manufacture.

1942 A team headed by Italian physicist Enrico Fermi creates an experimental nuclear reactor, a controlled fission chain reaction in a pile of uranium and graphite, at the University of Chicago. Based on his successful results, the Manhattan Project uses his method to split uranium atoms to create plutonium, which is used in the bomb dropped on Nagasaki, Japan.

1942 German-American biologist Ernst Mayr furthers Darwin's ideas on evolution, proposing that new species originate when a subpopulation of an ancestral species becomes geographically separated. Mutations and natural selection lead the subpopulation to develop new traits, and eventually this genetic drift leads to a group of individuals that can only breed among themselves, which Mayr defines as a new species.

1943 The first full-scale nuclear reactor is activated at the Oak Ridge National Laboratory in Tennessee, and production begins of the enriched uranium (pictured) that will be used in the bomb dropped on Hiroshima, Japan.

The Manhattan Project

The Manhattan Project is a massive scientific effort conducted under the authority of the U.S. Army Corps of Engineers, which costs $2.2 billion and employs over 130,000 people at more than 14 sites around the country. Based on fears that Germany is developing nuclear weapons, the U.S. government sets up its own secret project and hires the best physicists in the country, including some who had emigrated from Europe due to anti-Semitic policies, to quickly advance the science of nuclear weaponry.

1944 At the Rockefeller Institute for Medical Research in New York City, geneticists Oswald Avery, Colin MacLeod, and Maclyn McCarty determine that deoxyribonucleic acid, or DNA, is the hereditary material carried in genes and chromosomes. Researchers begin to understand that DNA provides both the blueprint and the operating instructions for most forms of organic life, but its structure still isn't understood.

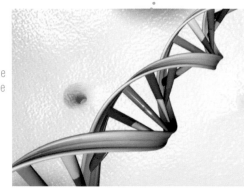

1945 The Manhattan Project, under the leadership of American physicist J. Robert Oppenheimer, succeeds in producing three atomic bombs. The first nuclear device is tested in July in New Mexico, and in August atomic bombs are dropped on the Japanese cities of Hiroshima and Nagasaki. Many of the scientists who worked on the project later speak out vehemently against the use of nuclear weapons.

1945 Fluoridation of the water supply to prevent dental decay is introduced in the United States. The fluoride ions are believed to strengthen tooth enamel, and the distribution of treated water through municipal water supplies is expected to provide wide public benefits. But some questions have persisted about its safety.

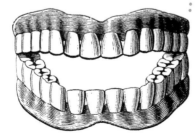

1945 Austrian zoologist Karl von Frisch finishes his study of honeybee communication, interpreting the complex dance a worker bee does to convey information about the distance and direction of food. The so-called "waggle dance" is a figure-eight maneuver a worker bee performs when he returns to the hive. Von Frisch's work is a significant step in ethology, the scientific study of animal behavior.

The Human Toll

The first military use of the atomic bomb occurs on August 6, 1945, when the bomb code-named Little Boy is dropped on Hiroshima. Its shockwaves destroy more than 80 percent of the city's buildings, and the intense heat and radiation kill between 70,000 and 80,000 people immediately. By the end of 1945, thousands more die of radiation poisoning, characterized by massive internal bleeding and the breakdown of tissue in the gastric and intestinal tracts. Another bomb, code-named Fat Man, is dropped on Nagasaki on August 9. More than 40,000 people die instantly.

Post–World War II Science: 1946 to 2006

As the world rebuilt following World War II, faith in science and in human ingenuity was high. After the destruction of Hiroshima and Nagasaki, many people looked forward to a prosperous peacetime, helped by "man's new servant, the friendly atom," as one National Geographic article put it. The military dreamed of nuclear-powered airplanes, while the chairman of the U.S. Atomic Energy Commission predicted that "our children will enjoy electrical energy in their homes too cheap to meter." In the coming decades, scientists would realize that the first proposition was technologically impractical, and that nuclear power was considerably more complicated than expected.

American scientists complacently assumed that they led the world in scientific progress until October 1957, when the Soviet Union launched the first artificial satellite into orbit, Sputnik I. "In the eyes of the world, first in space means first, period," wrote Lyndon Johnson. To regain an edge, American leaders put a new emphasis on math and science education in schools, and poured money into research programs. Then, in 1961, President John Kennedy requested billions of dollars from Congress to support the Apollo Project, which aimed to land an American on the moon by the end of the decade. While Kennedy didn't live to see the success of the undertaking, the project captured the public imagination at the height of the Cold War and showed the potential of focused and cooperative research.

The race to understand the structure of DNA, the chemical compound that carries genetic material, was less cooperative and more competitive, but the results were just as astounding. In the early 1950s, several teams in the United States and England were hard at work; they already knew that the molecule was responsible for passing down hereditary characteristics from parents to their offspring, but didn't know what it looked like or how it functioned. The team of James Watson and Francis Crick, working in Cambridge, England, solved the problem in 1953. The molecule was a double helix, they explained, and looked like a ladder with many rungs that had been twisted into a spiral form. The rungs were pairs of smaller chemical compounds called nucleotides, and the sequence of those nucleotides contained coded instructions on how to make various proteins, the building blocks of life.

Watson and Crick's discovery gave scientists a larger challenge, which made molecular biology the most exciting scientific field of the latter 20th century: to crack the code, and learn which sequences of DNA (also called genes) corresponded to which characteristics. Over the following decades, medical researchers sought to understand the genetic causes of hereditary diseases, and soon began to wonder if those diseases could be corrected by intervention at the genetic level. Genetic engineering, the manipulation of DNA, began in earnest in 1973, when three scientists working in California figured out how to transfer a gene from one species to another. The work was heralded as a breakthrough, but also raised concerns over the degree to which scientists were meddling with nature.

Indeed, by the 1970s public enthusiasm for scientific research was inhibited by a growing unease about possible unintended consequences. In 1962, the American biologist Rachel Carson gave voice to the fledgling environmental movement, calling public attention to the industrial chemicals, insecticides, and pesticides that were contaminating the water and air. People lost faith in the old 1950s slogan, "better living through chemistry," and began demanding that governments study and regulate the use of potentially toxic chemicals. Earth scientists soon realized that human activities were having an unprecedented effect on the atmosphere and climate. The industrialized world's reliance on burning fossil fuels for energy was increasing the level of carbon dioxide in the air, which created a sort of thermal blanket that trapped heat in the atmosphere. It was the largest example to date of how human activities, unnoticed and unchecked, could alter the delicate balance of life on Earth. Although atmospheric researchers noticed the global warming trend in the 1960s, policy makers and scientists are still debating how humans can best adapt to reduce our impact.

Research surged forward in many fields over the last half-century, and many of these advances were enabled by new computer technology. During World War II, an American military laboratory built one of the first computers to calculate missile trajectories; it weighed 30 tons and took weeks of programming to set up for a new set of calculations. It wasn't until the late 1970s that computers began to move from laboratory and industrial settings to offices and homes. Personal computers proliferated and improved rapidly in the coming decades; meanwhile, a handful of universities were testing an early version of the Internet developed by the U.S. Department of Defense.

One project made possible by computer technology was the Human Genome Project, a sprawling effort conducted by geneticists in the United States, China, Japan, France, Germany, and the United Kingdom. Beginning in 1990, the researchers worked to determine the entire sequence of nucleotides in human DNA, and to identify the genes within the long sequence, those stretches of nucleotides that code for protein production. Fragments of DNA were subject to chemical analysis, and powerful computers pieced the fragments together—the technology advanced so rapidly that it took four years to sequence the first billion nucleotides, but only four months to sequence the second billion.

The Human Genome Project was carried out with strong government support, but truly fueled by the curiosity of hundreds of scientists who were determined to gain an understanding of life at the most basic level. Human beings have displayed the same curiosity since our ancestors huddled in a cave over a million years ago, and poked at a coal until it burst into flame. Looking to the future, it seems certain that human ingenuity will lead us to places we can't even imagine.

1946 While studying the *E. coli* bacterium, American geneticists Joshua Lederberg and Edward Tatum discover "sex" in bacteria — that bacteria can exchange genetic material through process of conjugation. This transfer of genes can benefit the recipient bacterium by, for instance, making it resistant to antibiotics.

1947 A team of 3 American physicists at Bell Labs builds the first transistor, a tiny, solid-state device in which the electrical signals flow through solid pieces of a semi-conductor material. An improvement over vacuum tubes, in which the electrical signals pass through elements in the heated tube, because vacuum tubes can burn out or break and require more power. Marks the foundation of modern electronics. Transistors are quickly adopted in radios, amplifiers, and other electric devices, and eventually used in computer microchips.

Bell Labs

The research and development arm of the Bell Telephone Company develops a range of revolutionary technologies in the 20th century. Shortly after its founding in the 1920s, researchers demonstrate a long-distance television transmission. The lab keeps up with the times, working on transistors in the 1940s, and lasers in the 1960s. In the 1970s, researchers develop several computer programming languages. Today, researchers are studying nanotechnology and a new form of plastic transistors.

1947 American pilot and WWII flying ace Chuck Yeager is the first to break the sound barrier, flying an experimental craft, the rocket-powered Bell X-1. By flying faster than the speed of sound, which is approximately 1,087 feet per second, he creates shock waves that cause a sonic boom heard by observers.

1948 American chemist Linus Pauling shows that proteins like hemoglobin, the protein in red blood cells that carries oxygen, are made of amino acids arranged in a helical pattern, like a spiral staircase. Proposes that DNA, the basic unit of genetics, may have a similar structure.

1948 American scientists add a small rocket to the top of a captured German rocket, creating the first multi-stage rocket. By jettisoning the German rocket after expending its fuel, the lighter, smaller rocket reaches a greater height—79 miles above the ground, well beyond the atmosphere. This technology is later used to launch satellites and space shuttles (like the Saturn IB rocket, pictured).

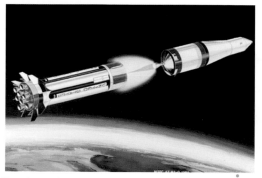

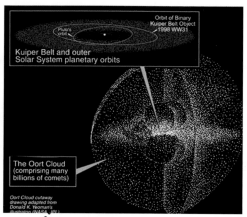

Kuiper Belt and outer Solar System planetary orbits

Orbit of Binary Kuiper Belt Object 1998 WW31

Pluto's orbit

The Oort Cloud (comprising many billions of comets)

Oort Cloud cutaway drawing adapted from Donald K. Yeoman's illustration (NASA-JPL)

1950 Dutch astronomer Jan Hendrik Oort proposes the existence of what will come to be called the Oort Cloud, a vast and distant cloud of rocky debris encircling the solar system. Explains the cloud as the source of most of the comets that enter the inner solar system, when the gravitational influence of a nearby star disrupts the cloud and sends comets in toward the sun.

1950

1949 During his treatment of burn victims in London during WWII, British zoologist Peter Medawar sees how often skin grafts are rejected when the attached skin comes from another person. After the war, his experiments show that a recipient's immune system often identifies the graft as foreign, causing the immune system to attack the new skin cells. In later skin and organ transplants, doctors use drugs to suppress this response.

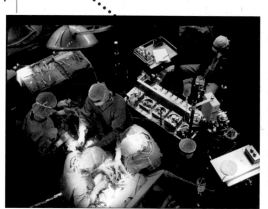

1951 American surgeon John Gibbon develops the heart-lung machine (shown, behind surgeons), a device that temporarily takes over the functions of circulating and oxygenating the blood while surgeons operate. Two years later he tries it on a patient while operating on her heart, and the operation is a success.

DNA Structure and Function

Knowledge of DNA's structure leads to an understanding of how the molecule can be reproduced, and therefore how the genetic information is replicated in new cells and passed down to offspring. The two strands of the double helix unzip from each other down the middle, and each side acts as a master copy to produce a new strand. Biological information is encoded in the strands by specific sequences of chemical compounds called nucleotides; the four different types are represented by the letters A, C, G, and T.

1951 American geneticist Barbara McClintock presents her discovery of "jumping genes" on the chromosomes of maize plants, genes that can move around on the chromosome. When a jumping gene lands in a certain places on the chromosome, it disrupts the activation of certain traits; McClintock studies the presence of pigment in the maize kernels through generations as a way of tracking the jumping genes.

1952 The first accident at a nuclear reactor occurs at Chalk River Laboratories in Canada, where a technician's error causes a partial meltdown in the nuclear core. The molten core penetrates its containment structure, and millions of gallons of radioactive water accumulate in the basement.

1952 One of the first sex-change operations is performed on George Jorgensen, who becomes a media sensation and is known to the world as Christine. Jorgensen, an American, travels to Denmark for surgery and hormone therapy, and becomes famous when a New York newspaper prints her story under the headline, "Ex-GI Becomes Blonde Beauty."

1953 Working in Cambridge, England, James Watson and Francis Crick develop the double-helix model for DNA, where the two helices are joined together by pairs of nucleotides that look like the rungs of a ladder. Their explanation is partially based on the work of Rosalind Franklin, whose x-ray diffraction photographs show a helical structure for DNA.

1953 At the University of Chicago, psychologists Nathaniel Kleitman and Eugene Aserinski open a "sleep lab" to study brain activity and eye movements during sleep. They find a link between a certain pattern of brain waves, rapid eye movements (REM), and dreaming, and show that sleepers have several short periods of so-called REM sleep during a night. When sleepers are awakened during a period of REM sleep, they usually report vivid dreams. The beginning of sleep research.

The Science of Dreaming

Continued sleep research and studies of brain activity during REM sleep have revealed some of the mechanics of sleep and dreaming, while leaving deeper questions still unanswered. Studies show that REM sleep is controlled by the brainstem, the part of the brain above the spinal cord, which turns off the neurons that control muscle movement to prevent dreamers from physically acting out their dreams. If sleepers are awakened repeatedly and thus deprived of REM sleep, they will spend more time in REM sleep during the next sleeping period to make up for the loss. This suggests that REM sleep and dreaming play a role in physical and mental health, but researchers don't yet understand that role.

1954 A team of Bell Lab scientists develops a workable photovoltaic cell, which can produce electric power from sunlight. Their silicon cell is 6% efficient, converting 6% of the free electrons it receives from the sun into usable electricity. Solar panels are first used to boost the power of satellites, and later installed on the roofs of homes and businesses to decrease energy costs.

1954 First nuclear power plant opens and connects to the electrical grid in Obninsk, USSR (control room, pictured). The reactor, which produces 5 megawatts of power, is used to power nearby businesses and residences, as well as for military research. It's considered a triumph for Soviet scientists during the Cold War's race for nuclear dominance.

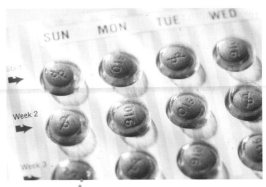

1955 American doctors John Rock and Gregory Pincus announce successful trials of the first birth control pills, which use the hormone progesterone to prevent women from ovulating. Because of strict laws against the distribution of all forms of contraceptive devices, the doctors were forced to do their work under the guise of a fertility study.

1956 First results published from the British Doctors Study, begun in 1951, to provide statistical proof that tobacco smoking increases the risk of lung cancer and heart disease. Refutes ads by the tobacco industry

that cite physical and mental benefits of smoking. The study follows 34,439 doctors over the course of 50 years, finishing in 2001.

1955 The tranquilizer Miltown is introduced, and is marketed to women as an anti-anxiety medication that can help them through their day. Famously called the "happy pill," it becomes the first bestselling psychotropic drug, and drug companies begin to search for more anti-anxiety and antidepressant drugs. By the 1970s, doctors realize Miltown is physically and psychologically addictive, and it is listed as a controlled substance.

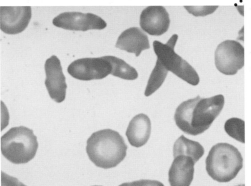

1956 British biologist Vernon Ingram demonstrates the genetic cause of sickle-cell anemia, a disease where misshapen red blood cells block blood vessels and lead to circulation problems and potential organ failure. Ingram shows that the mutation of just one "letter" in the DNA code for the hemoglobin gene causes the lifelong disorder.

The Space Race

With the United States and the Soviet Union locked into the Cold War, the launch of Sputnik is viewed as a sign of Soviet superiority in space technology. This is a blow to American pride and a national security concern, as U.S. leaders worry that Soviet scientists are also ahead on the technology of missiles that could carry an atomic bomb. To catch up, the U.S. spends millions on the space program, and changes school curricula to include more science and math.

1957 American surgeons announce the results of their studies of Henry M, an epileptic who had parts of his brain removed to stop his seizures. Studies show that the removal of his hippocampus prevents him from committing new events to long-term memory. An important step in the understanding of learning and memory formation.

1958 Scottish doctor Ian Donald is the first to use ultrasound imaging as a diagnostic tool. His device uses inaudible, high-pitched sound waves, and interprets the echoes that are bounced back as images. Donald and his colleagues soon use the technique to study fetal growth, and to diagnose complications during pregnancy.

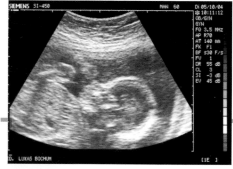

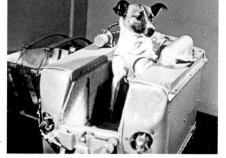

1957 In October, the Soviet Union launches Sputnik I, the first artificial satellite to orbit the earth. A few months later, it launches a second one containing a live dog. These early successes begin the "space race," when the Soviet Union and the United States compete to master the technology and engineering of spacecrafts.

1958 American astrophysicist Eugene Parker proposes that the sun produces a stream of electrically charged particles, plasma, that blows outward through the solar system, forming a "solar wind." Theory is based in part on German scientist Ludwig Biermann's observation that comets' tails always point away from the sun, as if blown back.

1959 In a landmark talk titled "There's Plenty of Room at the Bottom," American physicist Richard Feynman discusses the possibility of directly manipulating individual atoms. He envisions technical applications like atomic-scale data storage, computer circuitry, and machines. Widely viewed as launching the field of nanotechnology, which begins to develop in earnest in the 1980s.

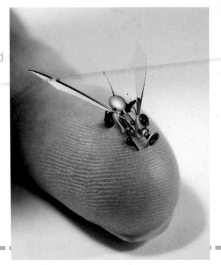

1959 The Antarctic Treaty is signed by 12 countries, including the U.S. and the Soviet Union, who promise to keep the continent free from military activities and to use it for scientific research. Research stations take advantage of the unique conditions around the South Pole to study geology, astronomy, and meteorology.

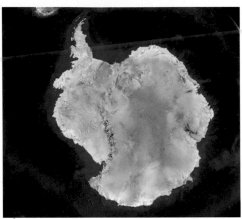

1960

1959 British anthropologists Louis and Mary Leakey, working in the Olduvai Gorge, Tanzania, find a skull dating from 1.75 million years ago. Their subsequent discoveries in the region convince anthropologists that humans originated in East Africa, and then spread through Asia, Europe, and beyond.

1960 The first laser, a narrow and intense beam of light, is developed by American physicist Theodore Maiman. While at first lasers are called "a solution looking for a problem," researchers soon find applications in consumer electronics, surgery, dermatology, and in industry, where lasers are used to cut steel and other metals.

1960 American geologist Harry Hess develops the theory of sea-floor spreading. Explains that hot magma rises to the surface at mid-ocean ridges where it pushes oceanic plates apart. The unified tectonic plate theory later builds on Hess's idea, adding that when a plate that has been pushed outward collides with another plate, they cause earthquakes, volcanoes, and the gradual creation of mountain ranges.

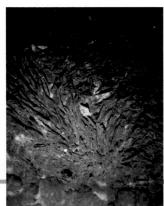

1961 Soviet cosmonaut Yuri Gagarin becomes the first human to travel in space. He orbits the earth once in less than 2 hours, and completes the mission safely, ejecting from the capsule as planned and parachuting to the ground.

1961 American biologist Leonard Hayflick shows that cells have a limited life span, and that they can divide only a certain number of times before dying. Human aging and mortality are related to this "Hayflick limit," which says human cells can divide about 50 times. This gives humans a potential lifespan of 120 years.

1961 In Europe, doctors realize that the sedative Thalidomide, given to pregnant women to control morning sickness, causes horrible birth defects. Signals the end of the public's unquestioning faith in "wonder drugs," and brings a new emphasis on drug testing and safety.

1962 American neuroscientist Roger Sperry shows that while the two hemispheres of the brain look identical, certain brain functions are performed primarily on one side or the other. Leads to the distinction between right- and left-brain functions: later studies show that the left side is primarily responsible for language and mental arithmetic, while the right side usually controls musical ability and spatial manipulation.

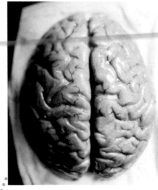

1962 Rachel Carson, an American biologist, publishes her book *Silent Spring*. Makes the general public aware of the introduction of chemical pesticides and industrial wastes into the ecosystem, and the harmful effects on wildlife. Credited with launching the environmental movement in the U.S., and encouraging the ban of pesticides like DDT.

1963 Dutch astronomer Maarten Schmidt recognizes that the "quasi-stellar radio sources" or "quasars" that radio telescopes pick up are extremely luminous, distant stars emitting massive amounts of electro magnetic energy. A quasar can radiate about 100,000 times the energy of the entire Milky Way galaxy, making them the brightest objects in the known universe.

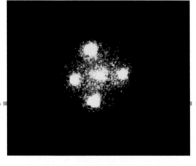

1963 Chinese embryologist Tong Dizhou produces the world's first cloned fish, inserting the DNA from a cell of a male carp into an egg from a female carp. However, he publishes his results in a Chinese journal that isn't translated into English, and his accomplishment goes unnoticed by the Western scientific establishment.

Mechanics of Cloning

Cloning can refer to several different processes that create an organism with identical genetic information to an existing organism. Identical twins are sometimes called natural clones, as they result from a single fertilized embryo that splits in two. The early efforts at cloning take an embryonic cell to copy, because it is already programmed to multiply and differentiate to form the different parts of an organism. Later, researchers learn how to take an adult cell and reprogram it to behave like an early embryonic cell.

1964 American biochemist Robert Holley works out the complete structure of transfer RNA, the molecule that helps build proteins based on instructions from DNA. By determining the nucleotides that it consists of, Holley shows how it attaches to and carries specific amino acids to protein-building sites within the cell.

1964 English biologist W. D. Hamilton demonstrates that there can be an evolutionary advantage to cooperation and altruism. Shows that an individual wasp sometimes engages in behavior that benefits its closely related kin, but not the individual itself. Hamilton shows that the altruistic action increases the chances that the individual's kin will survive and reproduce, so the family genes will be passed on. This runs counter to the traditional understanding that natural selection favors those individuals who best ensure their own survival and reproduction.

1965

1964 American physicist Murray Gell-Mann proposes that subatomic particles such as protons and neutrons are composed of still smaller particles called quarks. Gell-Mann's theory clarifies and simplifies particle physics, which explores the most fundamental level of matter. Other physicists later state that there are 6 different kinds of quarks that are held together by other particles called gluons.

1965 After breeding high-yield varieties of wheat in Mexico, American agricultural scientist Norman Borlaug starts the "green revolution" in India and Pakistan. Use of the new varieties nearly doubles the wheat yields in the two countries, at a time when they are afflicted by war and famine. Borlaug is credited with preventing over 1 billion people from starving on the Indian subcontinent.

1965 Working together at Bell Labs, American physicists Arno Penzias and Robert Wilson find the microwave background radiation left over from the immense explosion called the Big Bang, proving this theory of the universe's origins correct. Explain that the radiation emitted by the explosion was gradually dispersed throughout the

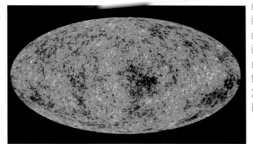

universe as it expanded outward, resulting in the background radiation at a temperature of 2.7 degrees Kelvin.

1966 The unmanned Soviet spacecraft Luna IX accomplishes the first soft landing on the moon in February, landing in the Ocean of Storms. The spacecraft sends photographs of the rocky, lunar terrain back to Earth, showing that the moon's surface is stable and solid enough to support a spacecraft.

1965 American paleobotanist Elso Barghoorn and geologist Stanley Tyler discover micro-fossils dating from 2 billion years ago in a quartz formation in western Ontario, Canada. The unicellular organisms from the Precambrian period increased the amount of oxygen in the air by photosynthesis, thus preparing the way for more complex, multicellular organisms.

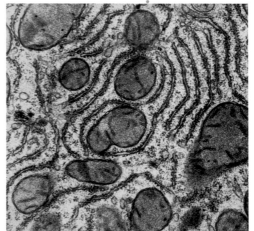

1966 Most plant and animal cells have structures in them called mitochondria, which convert food to energy (mitochondria in human liver, pictured). American biologist Lynn Margulis argues that mitochondria are descended from independent bacterial cells that were engulfed by larger cells, forging a mutually beneficial relationship. Her endosymbiotic theory of cell evolution is considered far-fetched at first, but is now generally accepted.

1967 Japanese meteorologist Syukuro Manabe and American Richard Wetherald, working together at the U.S. Weather Bureau, construct a computer model to study the effect of atmospheric carbon dioxide on global temperature. Warn that human activities that increase the amount of carbon dioxide in the air are causing a greenhouse effect that will raise global temperatures.

The Earth Warms

While the greenhouse effect is a natural phenomenon that keeps the Earth warm and hospitable to life, the term is usually used to refer to the amplified effect caused by the rapid increase of carbon dioxide and other "greenhouse gases" in the atmosphere. The gases are released when humans burn fossil fuels in power plants, factories, and cars, and they act as a thermal blanket that traps heat. Studies show that average global temperatures are projected to rise by 2 to 10 degrees Fahrenheit by 2100, leading to the rise of sea levels and altered weather patterns, with more frequent floods and droughts.

1968 Swiss microbiologist Werner Arber discovers that bacteria defend themselves against viruses by producing enzymes that cut the DNA of the viruses at particular points. Later, these restriction enzymes become a basic genetic engineering tool, used by researchers as "chemical knives" to cut pieces of DNA into defined fragments for closer study.

1967 South African surgeon Christiaan Barnard performs the first successful heart transplant in Cape Town, giving a 55-year-old man the heart

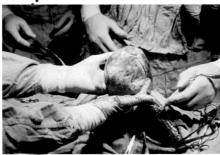

of a young woman killed in a car accident. The recipient lives for 18 days before dying of pneumonia; the immunosuppressant drugs he takes to prevent the rejection of the heart also prevent his immune system from fighting off the infection.

1969 Although Ohio's heavily industrial Cuyahoga River had caught fire several times in the past, a major river fire this year captures national attention when *Time* magazine describes a river that "oozes rather than flows." The fire, fed on oil slicks and debris, is quickly put out, but the public outcry encourages the passage of the federal Clean Water Act.

1969 The spacecraft Apollo 11 successfully lands on the moon on July 20th, and American astronaut Neil Armstrong becomes the first human to set foot on the moon's surface. As he steps from the ladder to the ground, he says, "That's one small step for man, one giant leap for mankind."

1970 Indian-born molecular biologist Hargobind Khorana announces the first complete synthesis of a gene, assembled directly from its component chemicals, called nucleotides. Artificially synthesized genes become indispensable tools in biotechnology, and today scientists can order strings of nucleotides from any number of companies. (Pictured, on right, with other Nobel prize winners from 1968.)

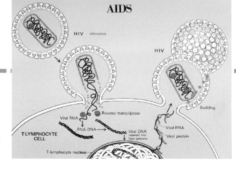

1969 Under the guidance of a U.S. Department of Defense research program, four computer systems at universities in California and Utah are connected into one network called ARPANET, a precursor to the Internet. Mainframe computers manage the network, and pass on data packets to individual computers that connect via modem. In 1971, the first email is sent.

1970 American geneticists Howard Temin and David Baltimore independently discover reverse transcriptase in viruses, an enzyme that copies genetic information from RNA onto DNA, the opposite of the normal direction. These "retroviruses" use reverse transcriptase to get their genetic material (in the form of RNA) into the DNA of the host organism. The host then replicates the virus's genetic material during the normal process of cell division, and the virus spreads. Later, several cancer-causing viruses and HIV (pictured) are determined to be retroviruses.

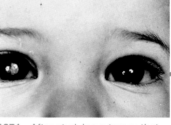

1971 After studying a tumor that occurs in the retina of the eye (pictured), American geneticist Alfred Knudson puts forth what will come to be called the Knudson hypothesis: that cancer can be caused by multiple mutations in a gene. Leads to the discovery of genes that control cancer, tumor suppressor genes.

1972   The use of the pesticide DDT is banned in the United States, following an outcry concerning its potential effects on the environment and human health. The chemical is particularly harmful to birds of prey like the bald eagle and peregrine falcon, as it causes the shells of the eggs they lay to weaken and become vulnerable. As a result, both the bald eagle and the peregrine falcon have a lower reproductive rate and are placed on the endangered species list in the 1970s.

1974   Three chemists, Mexican Mario Molina, American F. Sherwood Rowland, and Dutch Paul J. Crutzen, warn that the chemical compound chlorofluorocarbon (or CFC) commonly used in refrigeration and spray propellants is destroying the ozone layer in the atmosphere. Damage to the ozone layer, which shields the earth from ultraviolet radiation, would increase skin cancer and damage crops. An international treaty to ban production of CFCs takes effect 15 years later. (Ozone hole over Antarctica, shown.)

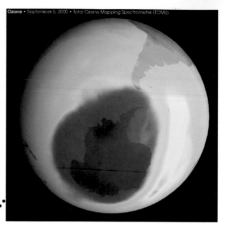

1973   American geneticists Paul Berg, Herbert Boyer, and Stanley Cohen show that DNA molecules can be cut with restriction enzymes, and the isolated gene can be transferred to another cell that will splice the gene into its existing DNA. This kind of DNA, which carries genes from more than one species, is called recombinant DNA. Marks the beginning of genetic engineering. (Magnification of human gene, pictured.)

Debate over Genetic Engineering

One year after the 1973 breakthrough, the National Academy of Sciences calls for a halt to genetic engineering research until its implications and potential risks are understood. Some scientists worry about moving DNA from one organism to another; the nightmare scenario is the creation of a hybrid virus that gets out of the lab and spreads through unprepared human populations. Molecular biologists soon come up with a self-governing system of guidelines and safety precautions; benefits from the continued research include the production of a genetically engineered bacterium that produces insulin to treat diabetes.

1974 British physicist Stephen Hawking proposes that some subatomic particles can escape from the edge of a black hole, taking away some of its mass and theoretically causing it to slowly evaporate over many billions of years. The existence of this Hawking radiation remains somewhat controversial.

1975 The first personal computer, the Altair 8800, is introduced in kit form in the United States at a price of about $400. The kit is an immediate success among hobbyists, and other companies scramble to enter the market, introducing features like the keyboard and the floppy disk for data storage. Two years later the Apple II personal computer (pictured) is the first to be sold in fully assembled form.

1975

Stephen Hawking

When Hawking was 21, he was diagnosed with the neurological disease known as Lou Gehrig's disease, and told that he had 3 years to live. The brilliant physicist, now 64, beat the odds on survival, but gradually became almost completely paralyzed. He now uses an electronic voice synthesizer to communicate, and operates a computer system with a "blink switch" to compose speeches and conduct his work. Without these technologies, Hawking would have had no way to share many of his discoveries and ideas with the public.

1975 New Zealand-born biologist Allan Wilson and his American student Mary-Claire King compare the DNA of humans and chimpanzee. Show that they're 98.5% identical, which means all the differences in body type, intelligence, and social behavior are coded in the remaining 1.5% of DNA. Also show that the 2 species split from a common ancestor about 5 million years ago.

1975 A book by American biologist E. O. Wilson presents the idea of sociobiology, the attempt to explain social behaviors by their genetic origins and evolutionary purposes. Shows, for example, that ants carry dead ants out of the nest not based on learned behavior, but because of a chemical trigger. Although most of his book deals with ant colonies, he devotes two controversial chapters to humans, stating that much of human behavior also has a genetic cause.

1977 A team of American oceanographers discovers a hydrothermal vent on the ocean floor near the Galapagos Islands that is the center of a complex ecosystem. Although life is usually very sparse at the frigid, black depths of the ocean, the mineral-rich and superheated water seeping from below the earth's crust at the vent feeds organisms that obtain energy from oxidizing sulfur compounds, in a process called chemosynthesis. These strange forms of life that can withstand intense heat, pressure, and toxic chemicals make scientists wonder if inhospitable planets harbor similar forms of life.

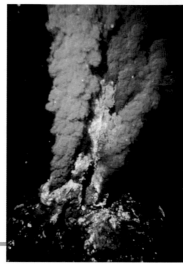

1976 American microbiologist J. Michael Bishop and Harold Varmus discover that oncogenes, genes that can cause tumor growth and cancer, are derived from genes found in normal cells. Show that these normal genes that regulate cell growth can be mutated by radiation, viruses, or exposure to some chemicals to become oncogenes (pictured, in cancer cell).

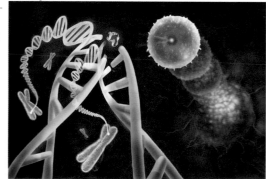

1977 Two men in New York City are diagnosed as having the then rare cancer Kaposi's sarcoma (pictured, in a highly magnified micrograph). In retrospect, doctors realize they are probably the earliest known patients with AIDS, or acquired immune deficiency syndrome. AIDS weakens the immune system and leaves patients more vulnerable to opportunistic infections like Kaposi's sarcoma, which is caused by a virus. The disease will not be officially recognized until 1981.

1977 After successful, worldwide vaccination campaigns, the last naturally occurring case of smallpox is reported in Somalia. Two years later, the World Health Organization declares that the disease has been eradicated in human populations, although cultures of the virus are kept by research institutions in the United States and the Soviet Union.

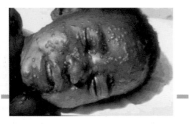

1978 The first human baby conceived outside a human body—called a "test tube baby" by the media—is born to Lesley Brown in England. In the technique of in vitro fertilization, ova are removed from the woman's ovaries and fertilized by sperm in a laboratory; the fertilized egg is then returned to the woman's uterus for the duration of the pregnancy. The procedure sparks controversy over the ethical implications of involving a doctor in conception.

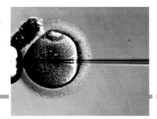

1978 Building on work by American chemist Paul Lauterbur, British physicist Peter Mansfield climbs into the first magnetic resonance imaging (MRI) scanner to have an image produced of his abdomen. In MRI, a non-invasive system used to examine the brain and internal organs, a patient is placed in a strong magnetic field and the variations produced by the spinning atomic nuclei in the patient's body are recorded.

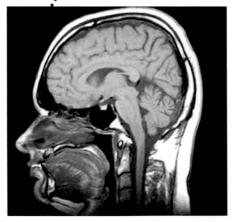

Fertility Treatments

Before the 1970s, the approximately 6 million Americans who were infertile had few options beyond adoption. In vitro fertilization brought new hope to people with a variety of medical problems, from damaged ovaries and fallopian tubes in women, to low sperm counts in men. In vitro fertilization had a low success rate at first—about 80 in vitro procedures were attempted before Lesley Brown succeeded in carrying her baby to term. The process was improved by placing the egg in the uterus sooner after fertilization, and by giving the mother the hormone progesterone, which keeps the uterus lining thick and ready for the egg's implantation.

1979 English biologist James Lovelock publishes a book arguing that the Earth can be viewed as a single living macro-organism based on the interdependence of organisms and their environment. Launches the modern field of ecology, and leads to a campaign to preserve biodiversity, or the range of species in an ecosystem, at a time when habitat loss in tropical rainforests is increasing extinctions.

1980 The Very Large Array, a radio astronomy observatory, begins operation in New Mexico. The array consists of 27 radio antennae each set in a dish 82 feet in diameter, which pick up radio wave emissions from the sun, planets in our solar system, and distant galaxies. The data from the antennae are coordinated to produce images of star formation, supernovae, and galaxies colliding.

Renewable Energy

As oil reserves are an exhaustible resource, and because burning fossil fuels contributes to air pollution and global warming, governments are increasingly turning to renewable energy sources, harnessing natural forces, such as sunshine, wind, geothermal heat, and ocean tides. "Wind farms" are being constructed on high hills and open plains, where the steady wind powers hundreds of windmills. Solar power is already in use both on the large-scale, for electricity generation in power plants, and on the small scale, with solar cells installed on the rooftops of homes and businesses to reduce electricity bills. Work remains to be done to increase the efficiency and lower the costs of these technologies, but experts expect them to play an important role in power generation in the coming centuries.

1980 A team led by American geologist Walter Alvarez and his father, physicist Luis Alvarez, discovers a thin layer of clay in limestone dating from the end of the Cretaceous period, the time at which the dinosaurs and many other animals went extinct. The clay is enriched with the rare metal iridium, leading the team to speculate that a giant body from space crashed into the earth at that time. This impact event would have caused clouds of dust that covered the earth, causing climate and environmental changes.

1981 Working at IBM's Zurich laboratory, German-born physicist Gerd Binnig and Swiss physicist Heinrich Rohrer invent the scanning tunneling microscope to view individual atoms. A very fine probe is moved over the sample and voltage causes electrons to jump (or "tunnel") from the probe's tip to the sample, creating a weak electric current. The current between the tip and the sample's surface varies according to the substance's composition and structure, and these variations are translated into an image.

Stem Cell Potential

While the medical benefits of stem cells are still far off, researchers believe that they have enormous potential for curing diseases. Embryonic stem cells can be grown into cells or tissue for medical therapies, but may trigger a reaction by the immune system. However, researchers are learning how to "reprogram" adult stem cells, found in the bone marrow, to behave like unspecialized embryonic stem cells. Hypothetically, a doctor could take stem cells from an adult patient with a failing liver, program them to grow into a new liver, and transplant it into the patient, without fear that the body's immune system would reject it as a foreign body.

1982 The American drug company Eli Lilly begins to sell Humulin to treat diabetes (company's insulin plant, pictured). This human insulin produced by bacteria is the first commercial product of genetic "engineering." The human genes responsible for producing insulin "are inserted into the DNA of a host bacteria cell, which then produces insulin. Previously, insulin was collected from the pancreases of cows, pigs, or fish, and caused occasional allergic reactions.

1981
Two separate teams of geneticists (one British, one American) develop embryonic stem cell lines, taken from mouse embryos, that continue to grow in the lab and can develop into practically any cell type. Stem cells are taken from embryos at an early phase of development, before they have differentiated into specific cell types, and are expected to have wide-ranging medical applications.

1981 Solar One, the world's first large-scale solar power station, is completed in southern California. A number of mirrors reflect the sun's light onto a black-colored receiver that collects heat, which is used to boil water to turn turbines. It generates up to 10 megawatts of electricity.

Prions

American biochemist Stanley Prusiner announces his discovery of prions, a class of infectious agents composed simply of a protein. The announcement is controversial because Prusiner shows that prions replicate without using DNA or RNA, a practice biologists previously thought impossible. One type of prion is later determined to cause mad cow disease by deforming normal proteins in the brain.

1983 French virologist Luc Montagnier and American Robert Gallo independently discover that AIDS is caused by HIV, the human immunodeficiency virus. Show that the retrovirus works by attacking white blood cells. Identification of the virus leads to a test for the disease, and helps researchers search for treatments.

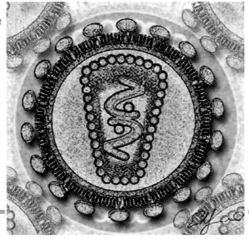

1984 British physicist Michael Green and American John Schwarz propose superstring theory, in which subatomic particles are not particles at all, but rather one-dimensional loops. These "superstrings" vibrate, and the type of vibration determines whether the string exhibits the properties of an electron, a quark, or another subatomic particle. Some physicists believe this to be a unifying theory that explains all forces and particles, while others think it's unverifiable nonsense.

Elk Cloner:
The program with a personality

It will get on all your disks
It will infiltrate your chips
Yes it's Cloner!

It will stick to you like glue
It will modify RAM too
Send in the Cloner!

1982 15-year-old American high school student Rich Skrenta creates what is thought to be the first computer virus, called "Elk Cloner." It affects the Apple operating system and is spread by floppy disk, automatically copying itself onto any uninfected disks that are inserted into an infected computer. Its effects are more annoying than damaging: it causes a self-congratulatory poem to pop up periodically on infected machines.

1985 British geneticist Alec Jeffreys announces his discovery of genetic fingerprinting: the identification of certain sequences of DNA that are unique to each person. The technique is soon used in forensic science, to match biological evidence with an individual, and in paternity testing.

1985 A team of chemists from England and the United States discover a new form of carbon molecule, composed of exactly 60 carbon atoms bound together in a sphere that looks like a soccer ball. The molecule is named buckminsterfullerene, because its shape recalls the geodesic domes built by architect Richard Buckminster Fuller, but it comes to be called a buckyball. It is expected to have applications in the emerging field of nanotechnology due to its heat resistance and superconductivity.

1986 Working at IBM's research lab in Switzerland, physicists Karl Müller and Johannes Bednorz invent a ceramic that is superconductive at 35 degrees Kelvin. This is the highest known temperature for superconductivity, the phenomenon characterized by zero electrical resistance and therefore zero loss of electricity during transmission. A step toward feasible and widespread technological applications, like superconducting wires that can efficiently transmit electricity over long distances.

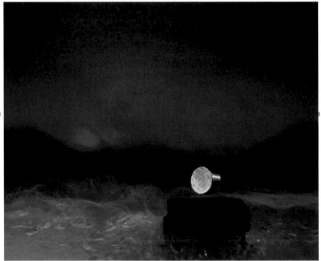

Nanotechnology

The broad term nanotechnology refers to the manipulation of matter on a nanometer scale, handling individual atoms or molecules. At that scale, matter exhibits different electromagnetic and optical properties that can be usefully applied in many technologies; researchers are now developing nanoscale computer chips and transistors to save space. Nanotechnology is increasingly used in consumer products, from cell phone displays to fabrics.

1986 American pharmacologist Louis Ignarro shows that the gas nitric oxide

NO

(NO) regulates blood pressure and blood flow to different organs, the first discovery of a gas used as a biological regulator. The gas is produced by cells in the arteries and spreads through cell membranes to the muscles that control the dilation and contraction of the arteries. The discovery has great potential applications in treating heart disease.

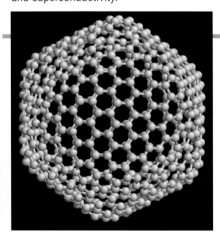

1986 American geneticist Louis Kunkel locates the gene that mutates to cause Duchenne muscular dystrophy, a progressive and fatal disease of muscle degeneration. The next year scientists show that the mutated gene doesn't produce the protein dystrophin, which is needed for healthy muscle fiber. With this understanding of the genetic cause and mechanism of the disease, researchers can begin to search for treatment.

Genetics in Anthropology

Previously, the study of human evolution and the migratory patterns of early humans was pieced together by archaeologists and anthropologists. They hunted for bone fragments and signs of habitation around the globe, and endeavored to put together the grand story of humankind. Biochemist Allan Wilson provided a new tool with his molecular clock, adding a genetic component to the study of human origins. His work was initially challenged by anthropologists who disagreed with his findings and believed that modern humans evolved much earlier, but in time his work is seen as definitive.

1987 The first drug is approved in the United States to treat AIDS and HIV infection, an antiretroviral drug usually called AZT. It slows the progress of the disease and the replication of HIV viruses, but can't destroy the infection all together. It is later used in combination with other anti-AIDS drugs in a "cocktail" that prevents the virus from proliferating. (Pictured: children in Kenya receiving liquid form of the drug.)

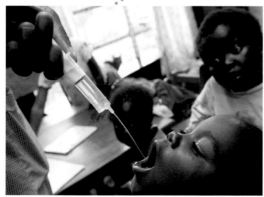

1987 A team of American geneticists led by New Zealand-born biochemist Allan Wilson (pictured, on right) studies the DNA found in mitochondria (mDNA), and use a "molecular clock" that assumes that mDNA evolves at a constant rate. They compare the mDNA of people across the globe, and by counting the differences they determine when different ethnicities diverged from a common ancestor. Conclude that all modern humans evolved from a population in Africa about 150,000 years ago.

1988 American doctor Graham Colditz announces that a study of 120,000 women shows that those who smoke half a pack of cigarettes a day are twice as likely to have strokes as non-smokers, and those who smoke 2 packs a day are six times as likely. Chemicals in tobacco are later shown to narrow blood vessels, increasing the chance of a blockage preventing blood flow to the brain.

1990 American geologists Alan Hildebrand and William Boynton find minerals that show the Caribbean region was the impact site of an enormous meteor 65 million years ago that probably caused the extinction of the dinosaurs. Later research on Mexico's Yucatan peninsula identifies the Chicxulub Crater as the site of the meteor's crash, which is about 100 miles in diameter.

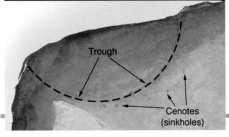

Trough

Cenotes (sinkholes)

1990

1988 The U.S. drug company Eli Lilly introduces the antidepressant drug Prozac, the first of a new class of psychotropic drugs that functions by regulating the neurotransmitter, serotonin. A low level of serotonin in the brain is one of the symptoms of clinical depression. Prozac becomes a bestseller thanks in part to an aggressive marketing campaign, leading some to question whether the drug is over-prescribed to patients looking for a quick fix for emotional pain.

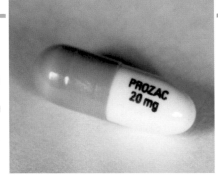

PROZAC 20 mg

1990 Human Genome Project is launched, an attempt to map out the entire DNA sequence carried on humans' 23 chromosome pairs, and to identify all the genes present. Collaborative effort is spearheaded by the U.S. National Institute of Health and Department of Energy, but geneticists from the United States, Europe, China, and Japan will all contribute.

CHEMISTRY BIOLOGY PHYSICS ETHICS INFORMATICS ENGINEERING

1990 First tests of gene therapy in humans conducted, for a genetic disease in which one nonfunctioning gene causes a completely inactive immune system. In the trial, American doctor W. French Anderson inserts a functional copy of the gene into the white blood cells of the patient, a four-year-old girl. The test is judged a qualified success: the girl has improved immune function but the injections have to be repeated over several years.

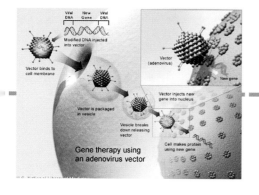

Gene therapy using an adenovirus vector

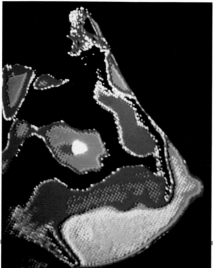

1990 American geneticist Mary-Claire King shows that certain variants of a particular gene indicate an increased risk of breast cancer, proof that the disease can have a hereditary component. The gene, BRCA1, is a tumor suppressor gene that regulates the division of cells that line the milk ducts in the breast.

1990 The Hubble telescope is launched into orbit around the Earth. After its mirror is fixed in 1993, it takes clear and beautiful images of nebulae, supernovae, and comets, among other objects. Its position outside the Earth's atmosphere allows researchers to observe very faint stars, and to study infrared and ultraviolet waves that are mostly absorbed by the atmosphere.

Leading Causes of Death in the U.S.

1. **Heart Disease**
   28% of deaths.
2. **Cancer**
   Lung cancer kills the most men and women; next is prostate cancer in men and breast cancer in women—23% of deaths.
3. **Stroke**
   6% of deaths
4. **Respiratory diseases**
   Including bronchitis and emphysema—5% of deaths.
5. **Accidents**
   Mostly car accidents and falls—5% of deaths.
6. **Diabetes**
   Possible complications include kidney failure—3% of deaths

1991 British geneticists Robin Lovell-Badge and Peter Goodfellow find the SRY gene on the Y chromosome that makes an embryo male and forms the testes. Show that when they insert the gene into a female mouse embryo (pictured) in the early stages of development, it then develops testes and male characteristics.

1991 At the European laboratory CERN, British computer scientist Tim Berners-Lee invents the web browser and puts up the first web site (his work station, shown). The site explains what the World Wide Web is, and offers instructions on how to set up web servers to store information, web sites to display information, and web browsers to search for information.

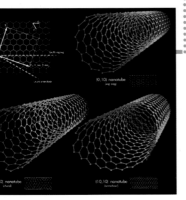

1991 Japanese physicist Sumio Iijima discovers a new form of carbon molecule, a hollow tube called a carbon nanotube that is in the same structural family as the previously discovered buckyball. The tiny tubes—10,000 times smaller than the width of a human hair—are one of the strongest materials ever discovered, due to the strong bonds between individual atoms. Expected to play an important role in forthcoming nanotechnology.

1992 A Japanese satellite named Geotail is launched to study the Earth's magnetosphere, the region of space controlled by the planet's magnetic field. Electrically conductive particles known as plasma are trapped by the magnetosphere forming several belts of radiation around the earth. The magnetosphere acts as a barrier to the solar wind, the stream of highly radioactive particles that flow outward from the sun.

1994 The first genetically modified food is approved for human consumption, the FlavrSavr tomato. It's engineered to resist rot and have a longer shelf life by the insertion of a gene that slows down the softening of cell walls as the fruit ripens. The announcement provokes an intense controversy over the potential benefits and risks of genetically modified organisms.

1995 Off the coast of Bermuda, American oceanographers launch the Autonomous Benthic Explorer, a 6-foot long underwater robot that prowls the sea bottom according to pre-programmed instructions. The robot collects samples of the ocean floor at depths of up to 16,000 feet, and maps water temperature to help researchers locate thermal vents.

1994 Fragments of the comet Shoemaker-Levy 9 collide with the planet Jupiter. The incident is predicted by the astronomers who discovered the comet, Canadian David Levy and husband and wife team Eugene and Carolyn Shoemaker, and the crash of the comet fragments and the resultant fireballs are closely observed. Scientists think the gravitational pull of the massive planet causes many small comets and asteroids to collide with its surface, and prevents more impacts on the Earth.

1995 The U.S. spacecraft Galileo goes into orbit around Jupiter, studying both the planet and the four large Jovian moons (the moon Callisto, shown). Investigations of the moon Io suggest that its surface is covered by a volcanic crust, with a mantle of partially melted rock below. Studies of the moon Europa indicate the presence of a salt water ocean beneath the thick icy crust, suggesting the possibility of primitive life forms.

1995 Controversial American psychologist J. Michael Bailey studies rates of homosexuality in both identical and fraternal male twins, and says that results show a genetic component or influence on sexual orientation. With 52 percent of identical twins, who are genetically identical, both men are homosexual, which is the case in only 22 percent of fraternal twins. Debate and research continue over the biological and environmental factors that determine sexual orientation.

1996 Deep Blue, a chess-playing computer system developed by IBM, wins a game against the reigning world champion, Garry Kasparov, for the first time. The computer is capable of evaluating 100,000,000 positions per second. But Kasparov goes on to win three games and draw two, winning the six-game series by a score of 4–2.

1996 The cloned sheep named Dolly is born in Scotland. While scientists had been cloning animals for some years by copying DNA from an embryonic cell, Dolly is the first animal whose DNA is taken from an adult cell—an udder cell. The DNA is inserted into an ovum that has had its nucleus removed, causing the cell to revert to an embryonic state and begin dividing.

1996 A carbon sequestration test project begins off the coast of Norway, with over 2,500 tons of carbon dioxide injected daily into a rock formation deep below the seabed. Researchers view the technique as a way to reduce the impact of global warming by removing excess CO_2 from the atmosphere and storing it below ground, where they predict it will stay for thousands of years.

CO_2

Cloning Concerns

The announcement of Dolly's birth ignites a firestorm of debate, as people worry that human cloning can't be far behind. Some simply find the cloning process unnatural, while others fear that humans will be cloned to create organs for transplant, or out of a misguided search for immortality. While, theoretically, a clone made from an adult human would produce a genetically identical person, environmental factors would presumably make the clone a different individual. In 1997, the U.S. government bans federal funding for research on human cloning.

1997 A strain of the bacterium *Staphylococcus aureus*, which can cause ailments ranging from skin infections to meningitis, is found to be resistant to a third class of antibiotics. Most strains are already resistant to two common types of antibiotics, leading doctors to call it a "superbug." Antibiotic resistance occurs as a result of natural selection, as bacteria with mutations that help them survive the antibiotic treatment pass on the trait to the next generation.

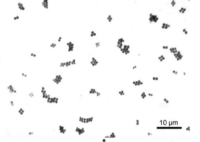

10 µm

Controversy over Stem Cells

The embryonic stem cells James Thomson uses are taken from extra, fertilized embryos produced during in vitro fertilization, which would otherwise have been destroyed. The embryos are at the blastocyst stage about one week after fertilization, when they consist of a mass of 50 to 150 cells. Ethical concerns are raised, because the creation of a stem cell line for research destroys the embryo and its potential for life. In 2001, the U.S. government decides to fund research using existing cell lines, not research that requires the destruction of a new embryo.

1997 Opening of the Polk Power Station in Florida, a new, highly efficient type of coal-fired power plant that doesn't emit the soot and gaseous pollution usually associated with coal power plants. Rather than burning the coal directly, the facility breaks it down into its basic chemical components, and produces a purified gas that is burned in the turbines. Seen as a model for how to use the Earth's vast coal deposits without causing excessive air pollution.

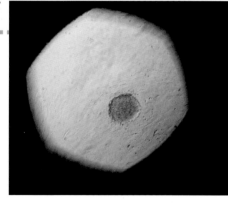

1998 American biologists James Thomson and John Gearhart independently report their successful isolation of a human embryonic stem cell, taken from frozen human embryos. From a single cell, researchers can create a "stem cell line," or a collection of constantly dividing cells. The development sets off a flurry of research into new treatments for human diseases, but also ignites a political and moral debate over the use of human embryos.

107

1998 The U.S. spacecraft Lunar Prospector orbits the moon, and discovers a high concentration of hydrogen at the north and south lunar poles. They are thought to indicate the presence of permanently frozen water ice in the polar craters, which was likely deposited there by comets and meteorites composed partly of ice. The discovery encourages those who dream of a settlement on the moon.

1999 In the Chinese province of Liaoning, a team of paleontologists led by Xiao-Chun Wu discovers a fossil of a feathered dinosaur, encased in volcanic ash 120 million years ago. The meat-eating dinosaur called *Sinornithosaurus* is thought to have had a downy coat of "proto-feathers" that were used for insulation, not flight. Supports the theory that birds evolved from a branch of dinosaurs.

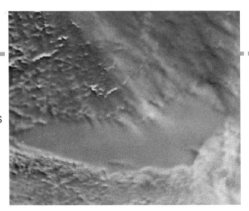

1998 Demonstration projects begin in Vancouver and Chicago of buses powered by a hydrogen fuel cell, built by the Canadian company Ballard Power Systems. In a fuel cell, hydrogen and oxygen react to create electricity, with water vapor the only byproduct. Research continues to make the technology cheaper and more viable, in hopes of creating an automobile that doesn't emit harmful gases.

1999 An international team of geologists announces discoveries from a Russian research station in Antarctica, where they drilled into the thick ice of Lake Vostok (pictured) and reached ice that formed 420,000 years ago. The ice core sample yields information on the Earth's climate, as the ice contains air bubbles that formed as each layer froze. Analysis of the gases in the bubbles shows that carbon dioxide levels are higher now than ever before, indicating that human activity is responsible for its accumulation and the greenhouse effect.

2000 The international effort known as the Human Genome Project announces a "working draft" of the human genome, sequenced down to the four nucleotides that are the chemical code of DNA. The working draft sequences about 90% of the 3 billion nucleotides in the genome. The mapping of the genome also makes it easier to locate genes, the stretches of DNA that code for making proteins. Researchers estimate that there are 20,000 to 25,000 genes in the human genome, lower than earlier estimates.

2000 Independent teams of English and American biologists show that adult stem cells taken from human bone marrow (pictured) can form into cells that make up liver tissue. Previously, researchers thought that adult cells couldn't be "reprogrammed," and that cells taken from the bone marrow could only produce more bone marrow cells. Raises the possibility of using adult stem cells for regenerative medicine, thus avoiding the controversial use of embryonic stem cells.

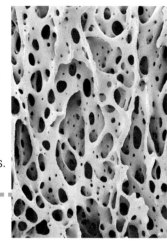

2000 A team of American engineers makes strides in organic electronics, in which carbon-based polymers are altered to become electrically conductive, replacing inorganic conductors like copper or silicon. These new organic transistors are cheaper to produce, flexible, and can be printed onto substances like paper, plastic, and cloth. Expected applications include flat panel displays and electronic tags to keep track of consumer goods.

EPSON
Organic Light Emitting Display

Mapping the Genome

The "genome" is the complete sequence of genetic information for an organism, the master blueprint that determines form and function. The $3 billion project to map the human genome's complexities is a resounding success, with a rough draft finished two years earlier than expected due to advances in the sequencing technology. Once researchers locate the genes in long strands of DNA, they can develop tests for diseases caused by the mutation of a particular gene, and can work toward medical treatments that target that gene.

Biological Warfare

The use of disease-causing organisms as weapons goes back nearly as far as war itself. In the 6th century B.C., the Assyrian army poisoned enemy wells with a dangerous fungus. During the Middle Ages, an army laying siege to a city sometimes catapulted the bodies of bubonic plague victims over the walls to infect the citizens. World War II brought a new level of sophistication, with labs in the U.S. and the U.K. developing highly infectious strains of anthrax and botulinum. In 1972, the Biological Weapons Convention banned the production of all weapons-grade biological agents. Their use is problematic in war, because diseases take some time to spread through enemy ranks and might also spread to allies. The most effective use of biological weapons, the decimation of a civilian population, is considered morally reprehensible.

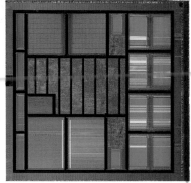

2001 A team of American chemists assembles molecules into a basic circuit, and show that its conducting potential can be switched on and off. Raises hope for advancements in nanoelectronics, where integrated circuits could be built on the molecular scale to make tiny, powerful, efficient computer chips.

2001 Letters containing anthrax bacteria (pictured) are mailed to the offices of several U.S. senators, newspapers and television stations in a bioterrorism attack that infects 22 people and kills 5. Anthrax, which is highly lethal when the spores are inhaled, has been studied by many countries as a potential biological weapon. The 2001 case remains unsolved.

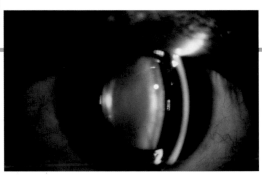

2001 59-year-old Robert Tools receives the first self-contained artificial heart in a surgery performed by doctors at the University of Louisville, but dies after five months of a stroke. Since then, doctors have increased the record to 17 months.

2002 American neuroscientist David Berson discovers a new type of cell in the eye that responds to light, and sends an electrical signal to the brain to govern the body's daily rhythms and internal clock. The cells, called retinal ganglion cells, help synchronize body functions such as sleep, body temperature, and hormone production to the rising and setting of the sun.

The Big Bang

In 2003 American astronomers announce findings from a spacecraft measuring microwave background radiation, or the radiant heat left over from the Big Bang that is thought to have marked the beginning of the universe. By studying the patterns of heat and the rate of expansion, astronomers determine the age of the universe to be 13.7 billion years old.

2003 26 years after its launch and 23 years after completing its fly-by mission past Jupiter and Saturn, the U.S. spacecraft Voyager 1 approaches the boundary of our solar system. The heliopause marks the edge of the Sun's magnetic field and the beginning of interstellar space. Scientists hope the Voyager 1, the most distant man-made object at over 8 billion miles from Earth, will continue to send back data from interstellar space.

2004 A team of Swiss and German physicists use a scanning microscope to study the leg of a spider, to determine how spiders cling to walls and walk upside-down along slick surfaces. They discover that each leg has tufts of hair, and each tuft is covered by hundreds of thousands of smaller hairs, called setules. These setules are tiny enough to exploit the attraction that exists between individual molecules to hold the spider's leg to the wall.

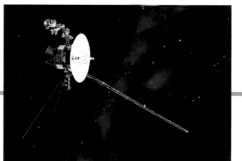

2003 A team of Swiss scientists shows that sunflowers planted in arsenic-contaminated soil effectively absorb the contaminant through their roots. While the plants are then contaminated, this clean-up method creates much less material to be disposed of than the usual technique of soil excavation. The discovery is one of many in the growing field of bioremediation, where plants and bacteria are used to clean up toxic waste.

2004 A team of Australian and Indonesian paleoanthropologists discovers fossils of a tiny species of human that lived 18,000 years ago. The 3-foot-tall humans lived on Flores Island in the Indonesian archipelago, and coexisted with modern humans for thousands of years before dying out. Researchers believe the species evolved from larger humans, but became smaller as an adaptation to the island's limited resources.

2004 Two robot rovers land successfully on Mars as part of the U.S. Mars Exploration Rover Mission. The rovers move over the terrain, explore several craters and hill ranges, and perform chemical analyses of the surface that suggest salty seas once existed on the planet, but have since evaporated.

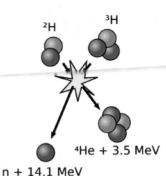

^2H ^3H

$^4He + 3.5$ MeV

n + 14.1 MeV

2005 A site in France is chosen for the first nuclear fusion power plant, an experimental program funded by an international consortium. Unlike standard nuclear fission power plants, the fusion plant is expected to produce very little radioactive waste, and is not thought to pose the risk of a meltdown due to the tiny amount of fuel contained in the reactor. The fusion reaction is the reaction carried out in the sun, where hydrogen isotopes combine to form a helium atom, releasing a burst of energy.

2005 A team of American geneticists use a tissue sample recovered from a woman killed by the 1918 flu and buried in the Alaskan permafrost to reconstruct the 1918 flu virus, and determine that it was a bird flu that jumped to humans. They then identify the mutations that made it particularly virulent, including an adaptation that speeded the virus's replication in host cells. This leads AIDS researchers to study the current avian flu virus in an effort to understand what mutations could make the virus even more dangerous to humans.

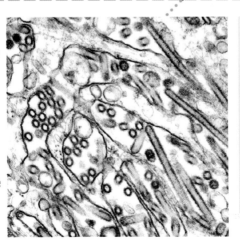

Avian Flu

Since 2003, a deadly strain of avian flu, a contagious disease that usually infects birds, has been diagnosed in 186 people in Asia and the Middle East, killing 105. As the number of human cases has increased rapidly, scientists warn of a potential pandemic. The virus has met the first condition for a human pandemic by adapting to a form that infects humans, but it has not yet met the second condition by mutating to a form that spreads easily from human to human.

2005 A team of British and German molecular biologists discover the gene in plants that initiates flowering. The gene, which is found in leaves, is activated by environmental factors such as increased temperature or sunlight, and begins producing a protein that causes undifferentiated stem cells at the shoot tips to form into flower buds. May allow scientists to breed new varieties that can flourish and flower in harsh conditions.

Nuclear Power

In the 1950s, a burst of enthusiasm for the new technology led U.S. leaders to predict that thousands of nuclear power plants would be built around the world, leading to plentiful electricity that could be distributed for free. A few major accidents dampened public enthusiasm, however, most notably the Three Mile Island accident in 1979 and the Chernobyl disaster in 1986, and environmentalists raised awareness of the waste produced, which remains radioactive for thousands of years. The U.S. now generates 20% of its electricity through nuclear power. Recently, as governments have become more concerned about the greenhouse gases produced by conventional power plants, new nuclear plants have been considered as a viable option. New, safer technology like nuclear fusion could help overcome strong public opposition.

2006 A team led by French-born glaciologist Eric Rignot shows that the glaciers and ice sheets of Greenland are melting faster than before, due to rising temperature. The amount of ice that melted into the sea doubled in a decade, which is seen as evidence that the effects of global warming are already being felt, and raises concerns about sea level rise.

2006 After 28 years of effort, the massive Chinese tree-planting project known as the Great Green Wall shows results in reversing desertification, in which overgrazing, poor agricultural practices, and the clear-cutting of forests spreads the northern desert. The Great Green Wall will eventually consist of a 2800-mile network of forest belts, from the edge of Beijing to Outer Mongolia, that will keep topsoil in place, and serve as a windbreak to prevent dust storms.

Photo Credits

Cover Image: DNA Double Helix: Ullstein Bild/The Granger Collection

FROM STONE TOOLS TO COPERNICUS

Stone Age Tool with Sharpened Edge: Erich Lessing/Art Resource, NY; **Grass Fire**: Photo by Robert Kerton/© CSIRO; **Paleolithic Lamp**: Photo by Jean Schormans: Réunion des Musées Nationaux/Art Resource, NY; **Weary Warrior Returning to Camp, Prehistoric Cave Painting**: Erich Lessing/Art Resource, NY; **Bone Needles**: Museo Ostiense, Ostia, Italy: Erich Lessing/Art Resource, NY; **Wheat**: Malcolm Paterson/© CSIRO; **El Caracol, Mayan Observatory**: SEF/Art Resource, NY; **Clay Counting Tokens**: Réunion des Musées Nationaux/Art Resource, NY; **Mesopotamia: Drinking Beer**: The Granger Collection, New York; **Mummy of Ankhef**: The Granger Collection, New York; **Painted Wooden Model of a Boat**: © British Museum/HIP/Art Resource, NY; **Two Daggers**: Erich Lessing/Art Resource, NY; **Two Egyptian Sundials**: Photo by Margarete Buesing: Bildarchiv Preussischer Kulturbesitz/Art Resource, NY; **Hittite Chariot**: The Granger Collection, New York; **Bronze Age Tools**: The Granger Collection, New York; **Irrigation Canals along the Nile River**: The Granger Collection, New York; **Imhotep**: The Granger Collection, New York; **The Great Pyramid, Giza**: The Granger Collection, New York; **Egypt: Weighing Heart**: The Granger Collection, New York; **Comet Hale-Bopp**: NASA Kennedy Space Center (NASA-KSC); **Stonehenge**: The Granger Collection, New York; **Mesopotamia: Farming**: The Granger Collection, New York; **Egyptian Water Clock**: The Granger Collection, New York; **Dagger, Chain and Other Equipment**: Erich Lessing/Art Resource, NY; **Glass-Blowing in Ancient Egypt**: The Granger Collection, New York; **Floor Heating in the Caldarium of the Varius Thermae-Eleuthera Agora**: Erich Lessing/Art Resource, NY; **Dried Fish**: Photo by Georges Jansoone/Wikipedia Commons; **Engravings of Stills, from the Works of Arabian Scholar, Geber**: The Granger Collection, New York; **Assyrian Aqueduct**: The Granger Collection, New York; **Eclipse of The Sun**: The Granger Collection, New York; **Natural Gas Flame**: © Kurt Stier/Corbis; **Pythagoras**: The Granger Collection, New York; **Cataract**: National Eye Institute, National Institutes of Health; **Anatomy: Brain**: The Granger Collection, New York; **Yangtze River**: Photo by Ian Sewell/Wikipedia Commons; **Hippocrates**: The Granger Collection, New York; **Aristotle**: The Granger Collection, New York; **Date Palm**: Photo by Stan Shebs/Wikipedia Commons; **Magnetite**: United States Mineral Survey and the Mineral Information Institute (USGS); **Page from Euclid's "Opus Elementorum . . . In Artem Geometriae"**: The Granger Collection, New York; **Maya Alphabet**: Wikipedia Commons; **Recycled Steel in Furnace**: © H. David Seawell/Corbis; **Archimedes and His Lever**: The Granger Collection, New York; **Snow Crystals**: The Granger Collection, New York; **Palestrina, Italy**: Wikipedia Commons; **Arabic Astrolabe from Iraq**: The Granger Collection, New York; **Caraway and Nut Grass, Painting from Byzantine Manuscript**: The Granger Collection, New York; **Katsusika Hokusai, Chrysanthemums and Bee," from the "Large Flowers Series**: Réunion des Musées Nationaux/Art Resource, NY; **Replica of Zhang Heng Seimosgraph**: Wikipedia Commons; **Zodiac Encircling a Geocentric (Ptolemaic) Universe**: The Granger Collection, New York; **Greek Physician Galen**: The Granger Collection, New York; **Pulley**: Wikipedia Commons; **Hypatia Murdered by the Followers of Cyril, Patriarch of Alexandria**: Image Select/Art Resource, NY; **Hagia Sophia**: The Granger Collection, New York; **Windmill and Waterwheel, French Manuscript Illumination**: The Granger Collection, New York; **The Diamond Sutra**: The Granger Collection, New York; **Alchemists at Work in a Laboratory, Title page from the Geber's *Philosophy of Alchemy***: The Granger Collection, New York; **A Page from Al-Khwarizmi's Book**: Wikipedia Commons; **"Joe and His Venerable Partner Talking Over Insurance**

Matters," Wood Engraving: The Granger Collection, New York; **Page from Manuscript of Hunayn's** *Treatise on the Eye*: The Granger Collection, New York; **Chinese Rockets**: The Granger Collection, New York; **"Midwives Attend a Childbirth," Colored Woodcut by Jost Amman**: The Granger Collection, New York; **Ancient Chinese Mariner's Compass**: The Granger Collection, New York; **Fire**: Wikipedia Commons; **Memory in the Third "Ventricle" of the Brain, from Albertus Magnus'** *Philosophia Naturalis*: The Granger Collection, New York; **Roger Bacon**: The Granger Collection, New York; **Mariner's Compass**: The Granger Collection, New York; **Portrait of Hugh de Provence (detail)**: Wikipedia Commons; **Rainbow**: Robert Kerton/© CSIRO; **Medical Teaching, Woodcut ,Title-Page from Mundinus'** *Anatomy*: The Granger Collection, New York; **American Athlete Preparing to Throw a Javelin**: The Granger Collection, New York; Ragusa: Wikipedia Commons; **Jan van Eyck, "The Arnolfini Wedding"**: Erich Lessing/Art Resource, NY; **Rafael, "The School of Athens"**:

Wikipedia Commons; **Divider, Circle, Triangle**: The Granger Collection, New York; **Leonardo Da Vinci, Embryological Drawings**: The Granger Collection, New York; **Childbirth**: © ROB & SAS/Corbis; **Detail from Map of the World and Account of Vespucci's Voyage, by Martin Waldseemuller**: The Granger Collection, New York; **Copernican Universe**: The Granger Collection, New York.

FROM LAUDANUM TO DARWIN
Opium Poppy: The Granger Collection, New York; **Niccoló Tartaglia**: The Granger Collection, New York; **Vesalius's Muscles**: The Granger Collection, New York; **Pope Paul III**: The Granger Collection, New York; **Camera Obscura**: The Granger Collection, New York; **Tycho Brahe**: The Granger Collection, New York; **Tycho Brahe's Instrument**: Wikipedia Commons; **Zacherais Janssen's Compound Microscope (6), Galileo's Microscope (7), Galileo's Thermoscope (18)**: The Granger Collection, New York; **Sir Francis Bacon**: The Granger Collection, New York; **Gilbert's "De Magnete"**: The

Granger Collection, New York; **Umbilical Cord**: Photo by Tristan Denyer; **Galileo Galilei**: The Granger Collection, New York; **Kepler's Illustration of the Planets**: The Granger Collection, New York; **Harvey's "On the Motion of the Heart and Blood"**: The Granger Collection, New York; **Burning Pile of Charcoal Briquets**: © Charles O'Rear/Corbis; **Adding Machine**: The Granger Collection, New York; **Robert-Fleury's "Galileo Before the Holy Office"**: The Granger Collection, New York; **Frans Hals's Portrait of René Descartes**: Wikipedia Commons; **Torricelli with Barometer**: The Granger Collection, New York; **Von Guericke's Air Pump**: The Granger Collection, New York; **Saturn**: The Granger Collection, New York; **Christiaan Huygens and the Pendulum Clock**: The Granger Collection, New York; **Red Blood Cells**: National Institutes of Health/Drs. Noguchi, Rodgers, and Schechter, NIDDK; **Boyle's "The Sceptical Chymist"**: The Granger Collection, New York; **Royal Society Coat-of-Arms**: © Kaihsu Tai; **Hooke's Structure of Cork from "Micrographia"**: The Granger

Collection, New York; **Margaret Canvendish**: © Bettmann/Corbis; **Fly Maggots and Pupae**: © George D. Lepp/Corbis; **Discarded Scallop Shells**: © CSIRO; **Resin from Pine Tree**: Wikipedia Commons; **Cassini**: The Granger Collection, New York; **Jupiter and Four Moons**: The Granger Collection, New York; **Microscopic Image of** *Salmonella enterica*: Wikipedia Commons; **Meadow, Sky, Sun and Clouds**: © Farhad Parsa/zefa/Corbis; **Isaac Newton and the Apple**: The Granger Collection, New York; **Grevillea lanigera, the Woolly Grevillea**: Photo by Karen Gough/© CSIRO; **Papin's Steam Digester**: The Granger Collection, New York; **Newton and a Prism**: The Granger Collection, New York; **Halley's Comet**: The Granger Collection, New York; **Mercury-in-Glass Thermometer**: © Science Museum, London/HIP/Art Resource, NYHIP/Art Resource, NY; **Mosquito**: The Granger Collection, New York; **Dentist and Pierrot**: Rue des Archives/The Granger Collection, New York; **Jethro Tull's Seed Drill**: The Granger Collection, New York; **Linnaeus' Drawing of** *Dryas *octopetala*: The Granger Collection, New York; **Albrecht von Haller in his Laboratory at Berne**: The Granger Collection, New York; **Hot Metalwork**: Wikipedia Commons; **Citron and Orange**: The Granger Collection, New York; **Benjamin Franklin**: The Granger Collection, New York; **Caduceus**: The Granger Collection, New York; **Lavoisier**: The Granger Collection, New York; **Arkwright's Spinning Frame**: The Granger Collection, New York; **Oxygen Molecule**: Wikipedia Commons; **Cuvier Studying Fossils**: The Granger Collection, New York; **Uranus**: NASA; **Montgolfier Balloon**: The Granger Collection, New York; **Lavoisier's Apparatus for Investigating Water**: The Granger Collection, New York; **Galvani Experiment**: The Granger Collection, New York; **Lavoisier**: The Granger Collection, New York; **Engraved Plate from Edward Jenner's "Inquiry into the Causes and Effects of the *Variolae Vaccinae*"**: The Granger Collection, New York; **Caroline L. Herschel**: The Granger Collection, New York; **Earth**: NASA; **Voltaic Pile**: The Granger Collection, New York; **Young's Wave Interference**: The Granger Collection, New York; **Plants Sprouting**: © CSIRO; **Robert Fulton's *Clermont***: The Granger Collection, New York; **New York Election, 1876**: The Granger Collection, New York; **Dalton's Chemical Elements**: The Granger Collection, New York; **Rush's Tranquilizing Chair**: The Granger Collection, New York; **Rocket Locomotive**: The Granger Collection, New York; **William Smith's Geological Map of Cumberland, 1824**: Wikipedia Commons; **Hans Christian Oersted**: The Granger Collection, New York; **World's First Photograph**: The Granger Collection, New York; **Tracheotomy, Neck**: Illustration by Jeremy Kemp; **Anatomy of the Brain**: The Granger Collection, New York; **Fertilized Egg**: Photo by Ederic Slater/© CSIRO; **Urine Samples in Beakers**: © Bill Varie/Corbis; **Frontispiece from Charles Lyell's *Principles of Geology*, 1857 Edition**: Wikipedia Commons; **Plant Cells and Nuclei**: © Clouds Hill Imaging Ltd./Corbis; **Surgery with Anesthesia**: The Granger Collection, New York; **Galapagos Finches**: The Granger Collection, New York; **Henry's Electric Motor**: The Granger Collection, New York; **Leaf**: PDPhoto.org; **Samuel Morse**: The Granger Collection, New York; **Daguerreotype**: The Granger Collection, New York; ***Staphylococcus aureus* Bacteria**: © Visuals Unlimited/Corbis; **Glacier**: The Granger Collection, New York; **Ambulance Racing from Hospital**: © ER Productions/Corbis; **Neptune**: NASA; **Man Producing Nitro-cellulose**: © Hulton-Deutsch Collection/Corbis; **The Faroe Islands**: © John Noble/Corbis; **James P. Joule**: The Granger Collection, New York; **Lord Kelvin**: The Granger Collection, New York; **Foucault's Pendulum**: Musée des Arts et Métiers, Paris, Photo by Hervé Marchebois; **Sir Edward Frankland**: The Granger Collection, New York; **Pollution Cartoon**: The Granger Collection, New York; **Vacuum Flasks**: The Granger Collection, New York; **Louis Pasteur in His Laboratory**: The Granger Collection, New York; **Mendel Diagram**: The Granger Collection, New York; **Charles Darwin as an Ape**, 1871: Wikipedia Commons; **Gustav Robert Kirchoff**: The Granger Collection, New York.

FROM INTERNAL COMBUSTION TO THE ATOMIC BOMB

Internal Combustion Engine: The Granger Collection, New York; **Kirchoff**: The Granger Collection, New York; **Brain of Leborgne**: Rue des Archives/The Granger Collection, New York; **London International Exhibition of 1862**: © Hulton-Deutsch Collection/Corbis; **Magnet from *Practical Physics*, 1914**: Wikipedia Commons; **White Noise**: Wikipedia Commons; **Joseph Lister's Carbolic Spray**: The Granger Collection, New York; **"Ominous Millennium"**: © Gabe Palmer/Corbis; **Periodic Table**: The Granger Collection, New York; **Peacock Feather**: The Granger Collection, New York; **Wilhelm Wundt**: The Granger Collection, New York; **First Telephone**: The Granger Collection, New York; **Louis Pasteur**: The Granger Collection, New York; **Edison's Light Bulb**: The Granger Collection, New York; **Illustration of Human Chromosomes**: Jane Ades, NHGRI; **Dam Across River, Appleton, Wisconsin**: Detroit Publishing Company, between 1880 and

1899/American Memory Collections, Library of Congress; **White Blood Cells**: CDC/Steven Glenn, Laboratory & Consultation Division; **Rhizobium Bacteria**: © CSIRO; **Telegraph Receiver**: Rue des Archives/The Granger Collection, New York; **Nerve Cells**: The Granger Collection, New York; **Asteroid Ida**: NASA Jet Propulsion Laboratory (NASA-JPL); **Chandra Captures Remnant of Star-Shattering Explosion**: NASA Marshall Space Flight Center (NASA-MSFC); **Lesions on Tobacco Leaf**: U.S. Department of Agriculture/Photo by Rob Flynn; **Friedrich Wilhelm Ostawald**: Wikipedia Commons; **Ribbon Diagram of an Enzyme**: Wikipedia Commons; **J. J. Thomson**: The Granger Collection, New York; **First Published X-Ray**: The Granger Collection, New York; **Marie Curie**: The Granger Collection, New York; **Black Salt-Marsh Mosquito**: The Granger Collection, New York; **Max Planck**: The Granger Collection, New York; **Uranium Atom**: The Granger Collection, New York; **Blood Transfusion**: The Granger Collection, New York; **Wright**

Brothers' Flight: The Granger Collection, New York; **Pancreas, from *Gray's Anatomy*, 1918**: Wikipedia Commons; **Ivan Pavlov**: The Granger Collection, New York; **Early Radio**: The Granger Collection, New York; **Fruit Flies**: The Granger Collection, New York; **Terraced Hot Spring Mineral Formations**: © Carol Cohen/Corbis; **Paul Ehrlich**: The Granger Collection, New York; **J. J. Thomson**: The Granger Collection, New York; **Rutherford's Apparatus**: The Granger Collection, New York; **Mercury**: Wikipedia Commons; **Stars in Orion**: NASA Jet Propulsion Laboratory (NASA-JPL); **Pangaea**: The Granger Collection, New York; **Bohr's Onion Atom**: The Granger Collection, New York; **Ford Assembly Line**: Ullstein Bild / The Granger Collection; **Alexander Graham Bell at the New York End of the First Long-Distance Telephone Call to Chicago in 1892**: The Granger Collection, New York; **Albert Einstein, Oil Painting**: The Granger Collection, New York; **Giant Molecules**: The Granger Collection, New York; **Hubble Deep Field View**: R. Wiliams (STScI), the HDF-S Team, and NASA; **Swarm of**

Ancient Stars: NASA STI (Scientific and Technical Information) Program; **Cyclone**: NASA STI (Scientific and Technical Information) Program; **Representation of Insulin Structure**: Health Sciences and Nutrition/© CSIRO; **Anatomy: Spinal Nerves**: The Granger Collection, New York; **First EEG**: Berger H. Über, Archives für Psychiatrie, 1929; **Andromeda Galaxy**: NASA Marshall Space Flight Center (NASA-MSFC); **Gorgonians, Ocean Floor**: CSIRO Marine Research/© CSIRO; **Lucy Australopithecus**: Rue des Archives/The Granger Collection, New York; **Genetic Mutations of American Cockroaches**: CDC/Andrew J. Brooks; **White Dwarf Star**: NASA, ESA, H. Bond (STScI) and M. Barstow (University of Leicester); **Werner Karl Heisenberg**: The Granger Collection, New York; **Charles Lindbergh and "The Spirit of St. Louis"**: The Granger Collection, New York; **Penicillia**: U.S. Department of Agriculture/Photo by Keith Weller; **Algae Cultures in Laboratory**: Robert Kerton/© CSIRO; **Planetary Nebula NGC 7293, also Known as the Helix Nebula**: NASA Jet Propulsion

Laboratory (NASA-JPL); **George Washington Bridge**: The Granger Collection, New York; **Ernest Orlando Lawrence in the Radiation Laboratory of the University of California with His Invention, the Cyclotron**: The Granger Collection, New York; **Helium Atom**: Wikipedia Commons; **Quality Inspectors Examine a Nylon Stocking**: © Hulton-Deutsch Collection/Corbis; *Streptococcus pneumoniae*: CDC/Dr. Mike Miller; **Brain that has Undergone a Leukotomy**: Wikipedia Commons; **Tasmanian Wolf, 1902**: Wikipedia Commons; **Alanine**: Wikipedia Commons; **Technetium**: Wikipedia Commons; **Professor Hans Krebs in His Laboratory**: © Bettmann/Corbis; *Rattus norvegicus*: NHGRI; **Lise Meitner in the Laboratory with Otto Hahn**: Ullstein Bild/The Granger Collection; **Krupp Steelworks at Essen, Germany**: The Granger Collection, New York; **Planet Earth with Core Exposed**: Wikipedia Commons; **Crop Duster**: © Ron Sanford/Corbis; **Eniac Computer**: Rue des Archives/The Granger Collection, New York; **Iodine and Alcohol**: CDC/Dr. Mae Melvin; **Vought-Sikorsky**

Centers for Disease Control and Prevention; **ICSI Sperm Injection into Ooctye**: Courtesy RWJMS IVF Laboratory; **MRI, Brain**: Wikipedia Commons; **Stream and Forest**: Wikipedia Commons; **Very Large Array at Socorro, New Mexico**: Photo by Hajor; **Father-and-Son Research Team Luis and Walter Alvarez**: © Roger Ressmeyer/Corbis; **Scanning Tunneling Microscope**: Agronne National Library/Photo by George Joch; **Mouse Embryonic Stem Cell**: National Science Foundation; **Solar Two Heliostat**: Wikipedia Commons; **Eli Lilly Insulin Plant**: © Patrick Bennett/Corbis; **AIDS Virus**: U.S. Department of Energy; **Physicist John Schwarz**: © Philippe Caron/Sygma/Corbis; **Reading Genetic Fingerprints**: © Tom & Dee Ann McCarthy/Corbis; **Buckminster-Fullerene**: Wikipedia Commons; **Meissner Effect**: Courtesy of Pacific Northwest National Laboratory; **Accessible Public Space for the Disabled**: © Ed Kashi/Corbis; **Kenya's Children Battle with AIDS**: © Radhika Chalasani/Corbis; **Berkeley Professors Vincent Sarich and Allan Wilson**: © Roger

Ressmeyer/Corbis; **Senior Woman**: © Benelux/zefa/Corbis; **Prozac Pill**: © William Whitehurst/Corbis; **Shaded Relief with Height as Color, Yucatan Peninsula**: NASA Jet Propulsion Laboratory (NASA-JPL); **Human Genome Project Logo**: U.S. Department of Energy Human Genome Project; **Gene Therapy**: National Institutes of Health, U.S. Department of Health and Human Services; **Mammogram Cancer Cell in Breast**: © Howard Sochurek/Corbis; **Hubble Space Telescope**: NASA; **Mouse Embryo at Two-Cell Stage**: © Clouds Hill Imaging Ltd./Corbis; **Carbon Nanotubes**: Created by Michael Ströck; **First Web Server**: CERN Museum, Switzerland; **Workers Prepare Geotail**: © Corbis; **Tomatoes**: U.S. Department of Agriculture/Photo by Scott Bauer; **Comet Shoemaker-Levy 9**: H.A. Weaver, T.E. Smith (Space Telescope Science Institute), and NASA; **WHOI Autonomous Underwater Vehicle**: National Oceanic and Atmospheric Administration/Department of Commerce; **Global Callisto in Color**: NASA Jet Propulsion Laboratory

(NASA-JPL); **Twins**: © Meeke/zefa/Corbis; **Dolly, the Cloned Sheep at Roslin Institute**: © Reuters/Corbis; **Chess Player Garry Kasparov in Match**: © Najlah Feanny/Corbis SABA; **Tampa Electric Power Company's Polk Power Station at Night**: U.S. Department of Energy; *Staphyloccus aureus*: CDC; **Embryonic Stem Cells Research**: © Kat Wade/San Francisco Chronicle/Corbis; **Lunar Prospector**: NASA; **Fuel Cell**: NASA; **Model of a Sinornithosaurus at the American Museum of Natural History**: © SETH WENIG/Reuters/Corbis; **Lake Vostok**: NASA Goddard Space Flight Center; **Image of Silhouettes with Sequence Colors**: National Institutes of Health, National Human Genome Research Institute/Photo by Jane Ades; **Bone Tissue**: © Micro Discovery/Corbis; **Seiko Epson Unveils the First 40-Inch OLED in Tokyo**: © Issei Kato/Reuters/Corbis; **Computer Chip**: Wikipedia Commons; **Anthrax Bacteria**: The Anthrax Vaccine Immunization Program (AVIP); **Artifical Heart**: National Institutes of Health, U.S. Department of Health and Human Services; **Normal Adult Lens**: National Eye

Institute, National Institutes of Health; **Sunflowers**: U.S. Department of Agriculture/Photo by Bruce Fritz; **Artist's Concept of Voyager**: NASA Jet Propulsion laboratory (NASA-JPL); **Long-Jawed Orb Weaver**: U.S. Department of Agriculture/Photo by Scott Bauer; **Artist's Drawing at the Australian Museum of a Newly Discovered Species**: © Peter Schouten/National Geographic Society/Reuters/Corbis; **Spirit Beholds Bumpy Boulder**: NASA/JPL-Caltech/Cornell/NMMNH; **Fusion**: Wikipedia Commons; **Avian Influenza**: CDC/C. Goldsmith, J. Katz and S. Zaki; **Clivia**: Wikipedia Commons; **Greenland**: Wikipedia Commons; **Bamboo Forest in China**: © Keren Su/Corbis.